Circling the Drain

Humorous Musings on Becoming a Mechanical Engineer

RICHARD E. KLEIN, PhD

BS Mechanical Engineering, Pennsylvania State University (1964); MS Mechanical Engineering, Pennsylvania State University (1965); PhD Mechanical Engineering, Purdue University, (1969). Dr. Klein is a life member in both ASME and IEEE.

Published by Dumb Dickie Press

ISBN-13: 978-1-0990-7264-2
ISBN-10: 1-0990-7264-6

Cover design by Ellen Meyer and Vicki Lesage

Disclaimer

This document was initially drafted in the 1980s. I recently retrieved it, dusted it off, and made some editorial updates. Many of the facts and statements are based on happenings some 35 years past. The numbers may have changed, but the depictions remain true. The picture of mechanical engineering painted in the 1980s hasn't changed much. This document has its sad aspects, but it is also uplifting and filled with joy and optimism. The joy and optimism come because light is allowed to shine. Light beats darkness every time.

Table of Contents

Introduction

I want to start by describing my upbringing and the odd circumstances that caused me to study mechanical engineering. I was raised in a working-class family. My mother had an eighth-grade education. My father did complete high school but spent his life as a machinist. There was no silver spoon in my mouth. The Great Economic Depression and then World War II caused many hardships. When peace and prosperity followed WWII, a higher standard of living also arrived.

As was the case for most draft-age males while the Cold War with the Soviets intensified, the question of Selective Service had to be dealt with. Considerations of college and entering the job market were secondary. I was physically fit, classified as 1-A. In my mind, I had what was then called a military obligation. I had to serve a minimum of six years: two years active duty along with four additional years in the reserves. An alternative option of six months active and then seven and a half years of reserve duty didn't appeal to me. Some draft-age men took college deferments. That route likewise didn't appeal to me. I wanted my military obligation done with. Behind me. And with the shortest path possible. I enlisted right away.

Other factors pointed me towards joining the military as

opposed to entering college. For one thing, I was fed up with school. I had my fill of diagraming sentences and conjugating verbs that nobody gave a hoot about. I could not care less about split-infinitives and past-participles.

Lack of money was another factor.

Most importantly, I lacked good enough grades to get accepted to college. My elementary school record was good, but upon entering high school, my grades plummeted. In my junior year, I took a part-time job in a gas station. My father urged me to get out of the house and "make a buck." Soon the part-time job expanded to become a forty-hour workweek. I worked four hours each day after school and then ten hours on both Saturdays and Sundays. I was smart enough to pass my exams without much studying, but doing homework got the short end.

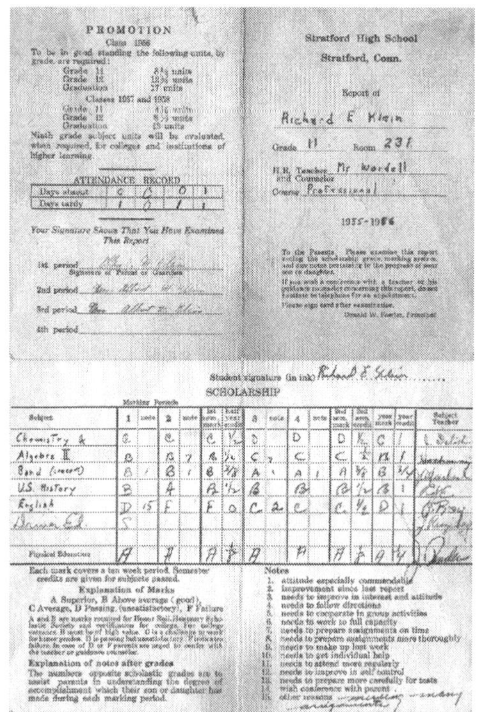

PROMOTION

Class 1956

To be in good standing the following units, by grade, are required:

Grade 11 — 8½ units
Grade 12 — 12½ units
Graduation — 27 units

Classes 1957 and 1958

Grade 11 — 4½ units
Grade 12 — 8½ units
Graduation — 15 units

Ninth grade subject units will be evaluated, when required, for colleges and institutions of higher learning.

ATTENDANCE RECORD

Days absent	0	0	0	1
Days tardy	1	0	1	1

Your Signature Shows That You Have Examined This Report

1st period [illegible]
Signature of Parent or Guardian
2nd period [illegible]
3rd period [illegible]
4th period

Stratford High School

Stratford, Conn.

Report of

Richard E Klein

Grade 11 Room 231

H.R. Teacher and Counselor Mr Wardell

Course Professional

1955-1956

To the Parents. Please examine this report noting the scholastic grades, marking system, and any notes pertaining to the progress of your son or daughter.

If you wish a conference with a teacher or his guidance counselor concerning this report, do not hesitate to telephone for an appointment.

Please sign card after examination.

Donald W. Fowler, Principal

Student signature (in ink) [illegible]

SCHOLARSHIP

Marking Periods

Subject	1	note	2	note	1st sem. mark	half year credit	3	note	4	note	2nd sem. mark	2nd sem. credit	year mark	year credit	Subject Teacher
Chemistry &	C		C		C	1/2	D		D		D	1/2	C	1	[illegible]
Algebra II	B		B	7	B	1/2	C	7	C		C	1/2	B	1	[illegible]
Band (concert)	B	1	B	1	B	3/8	A	1	A	1	A	3/8	B	3/4	[illegible]
U.S. History	B		A		B	1/2	B		B		B	1/2	B	1	[illegible]
English	D	15	F		F	0	C	2	C		C	1/2	D	1	[illegible]
Driver Ed.	S														[illegible]
Physical Education	A		A		A	1/8	A		A		A	1/8	A	1/4	[illegible]

Each mark covers a ten week period. Semester credits are given for subjects passed.

Explanation of Marks

A Superior, B Above average (good), C Average, D Passing (unsatisfactory), F Failure

A and B are marks required for Honor Roll, Honorary Scholastic Society and certification for college. For college entrance, B must be of high value. C is a challenge to work for higher grades. D is passing but unsatisfactory. F indicates failure. In case of D or F parents are urged to confer with the teacher or guidance counselor.

Explanation of notes after grades

The numbers opposite scholastic grades are to assist parents in understanding the degree of accomplishment which their son or daughter has made during each marking period.

Notes

1. attitude especially commendable
2. improvement since last report
3. needs to improve in interest and attitude
4. needs to follow directions
5. needs to cooperate in group activities
6. needs to work to full capacity
7. needs to prepare assignments on time
8. needs to prepare assignments more thoroughly
9. needs to make up lost work
10. needs to get individual help
11. needs to attend more regularly
12. needs to improve in self control
13. needs to prepare more carefully for tests
14. wish conference with parent
15. other reasons [illegible]

Report card from junior year of high school

I did graduate from high school, but with only mediocre grades. I was at the 59th percentile in class standing. Attending college wasn't in my crystal ball; I had written off the idea of going as I was geared up mentally to enter the military. I volunteered and joined the United States Army Enlisted Reserve on May 25, 1956, at age seventeen. I was still a junior in high school.

My strategy was to get in sooner and thus out sooner. My enlistment was a calculated move, based on the world as it was.

Unexpectedly, the world changed. President Dwight D. Eisenhower, in October of 1956, issued an executive order. It read simply: Anybody in the reserves who attended regularly would no longer be classified as 1-A. Prior to Eisenhower's executive order, my draft classification was 1-A, irrespective of my enlisted reserve status.

This unexpected twist in my draft status meant that I wouldn't be drafted. But with average grades, acceptance at most colleges was beyond my reach.

In the above photograph, taken in the summer of 1957, I was preparing to leave for summer camp at Camp Drum in upstate New York. I was eighteen at the time. I had just finished pulling the engine on my 1946 Ford coupe. In my forthcoming autobiography, *Dumb Dickie*, I recount the story behind that engine and how I had destroyed it. Note that in the photograph the cylinder head was already removed.

Fortunately for me, the family's church synod provided an avenue. The grade requirement for admission was merely placing in the upper half of the graduating class. My family was able to budget the modest cost for room, meals, books, and tuition, then at about $1,000 per year. Some of that money came from a small inheritance from my mother's parents. I applied to Grand View College in Des Moines, Iowa, and was accepted. Grand View was at that time a two-year junior college. Grand View "Junior" College served as a stepping stone that later got me into the mainstream of being a college student. My future now rested on my college grades, the grades I earned at Grand View College. My lackluster high school standing became irrelevant.

As a freshman at the small, church-supported junior college in Des Moines I was having a ball. Working forty hours each week was now behind me, at least for a few years. In 1957, the state of Iowa had not yet legislated daytime highway speed limits. The posted speed limit signs simply read "Reasonable and Proper." To me, 80 mph and even 100 mph seemed perfectly reasonable and proper. My '46 Ford coupe with its V-8 engine and triple Stromberg carburetors just purred along.

Along with college admission came the choice of a major, which would affect my future career path. I knew I liked mechanical things, most notably cars and car engines. In high school I had been known for having my head perpetually under the hood of my car.

Grand View College offered basic courses that prepared me to later select mechanical engineering for my major. I took the usual mathematics, physics, chemistry, drawing, and general classes. I even took a semester-long course on logarithms and how to use a slide rule. Those handy little things called calculators were yet to come.

Two years later, in the fall of 1959, I transferred to the University of Iowa. That was when I committed myself to becoming a mechanical engineer. My goal was to become an automobile designer.

While at Grand View College, the world and its ground rules suddenly changed. On October 4, 1957, the Soviet Union placed Sputnik into Earth's orbit. In Russian, the word *Sputnik* translates as wanderer or traveler. The United States was caught totally off guard. We didn't even know enough to be monitoring radio signals from overhead as we hadn't anticipated orbiting wanderers. A Russian news agency advised the United States to listen to a given frequency. Sputnik passed over our heartland, sounding a simple beeping signal.

The world wasn't ready for the Space Age, but no one could do anything to stop the Soviet satellite, which circled the globe and passed over the United States every 96 minutes. American society and education were jolted by Sputnik. America became obsessed with catching up with the Soviets. In 1961, President John F. Kennedy announced that the United States would place a man on the moon and return him safely to Earth within the decade. The space race was in full throttle mode.

As an engineering student, I was at ground zero. The post-Sputnik era in America's technical education spanned twelve years, from 1957 to 1969. The frenzy following Sputnik came to a screeching halt in 1969. Man's landing and walk on the moon, racial strife in America, and Vietnam War protests changed the landscape and our society.

Eleven years following Sputnik's launch, in 1968, I emerged from Purdue University with a doctorate in mechanical engineering. In September 1968, I became a new faculty member at the University of Illinois in Urbana-Champaign. My specialization was in feedback control systems. Said in other terms, I was a full-fledged rocket scientist. Guidance and control were my areas of expertise.

During my three decades at Illinois (1968-1998), I taught approximately 5,000 students at both undergraduate and graduate levels. During my tenure, The University of Illinois' Department of Mechanical Engineering was nominally ranked fourth in the nation. Two of the three schools ranked ahead of Illinois were privately endowed, including M.I.T. and Stanford University. Illinois was the leading university among the major land-grant universities.

With this background stated, I feel qualified to render some musings on mechanical engineering as a professional career choice.

The Origins of Mechanical Engineering

Since antiquity, mankind has functioned as *want-to-be* mechanical engineers. Stones were chipped to form tools. Tools of stone permitted other tools to be made. The development of tools led to yet other useful objects. Examples include the bow and arrow, the wheel and axle, the catapult, vessels for traveling on water, agriculture-related implements, textiles, and clothing. As society's needs increased, simple tools and machines evolved: the lever, the inclined plane, and the screw. The pyramids were built. Of course, one can argue that the pyramids were the result of civil engineers, but the advent of tools and simple machines was foundational. Warfare brought other innovations, including gunpowder and the ability for a projectile to penetrate armor.

The ability to harness energy from wind and falling water was a milestone. But the machines that harnessed this energy had a limitation: they were usually fixed in place as well as site-specific.

The early water wheels powered a grinding stone to grind grain.

The invention of the steam engine triggered a revolution. Steam engines could be placed where needed. The early steam engines had problems, however. The first problem was that the pressure vessel would at times rupture—thus injuring or even killing nearby people. The second problem concerned speed regulation.

Flyball governor on a steam engine

Over time and by trial and error, safer pressure vessels evolved. James Watt (1736-1819) addressed the speed regulation problem. Watt was familiar with the use of centrifugal weights, also known as flyballs, to regulate windmills and clock mechanisms. Watt's flyball governor application to steam engines caused a technological breakthrough. The industrial revolution shifted into high gear.

Of course, Leonardo DaVinci is considered to be one of the first mechanical engineers.

Leonardo DaVinci sketch of design for helicopter. Photo courtesy of Wikimedia Commons.

Sketch of bicycle design attributed to Leonardo DaVinci

His sketches depicted things unimaginable at the time—the bicycle and helicopter to name just a few. Although DaVinci was a revolutionary thinker, most of his ideas never saw creation in his time. The application of Watt's innovations, in contrast, jolted western civilization. The steam engine permitted mankind to harness energy abundant in nature, control it, and achieve portability.

Mechanical engineering as a profession evolved in large part because of the steam engine. Mechanical engineering had its roots in the work and accomplishments of common people trying to solve their problems. In stark contrast, numerous other professions evolved from educated people—people of wealth and class standing. Educated people used precision in language. Working class people, however, used everyday language and at times used it incorrectly.

Because the steam engine was so critical to the conception of modern mechanical engineering, two gene pools became co-mingled. One pool of genes came from steam—and everything about steam. That has been given a fancy name: thermodynamics. The second gene pool came out of mechanical design—the concept of the piston, the slider-crank, the flywheel, and the enclosure to contain and thus harness steam.

The two parents of mechanical engineering were thermodynamics and design of machine elements. And as in many accidental marriages, thermodynamics and mechanical design endured a long history of not getting along well. The marriage has never been harmonious or tranquil. Mechanical engineering still carries the scars of its illegitimate conception and birth. Although if its parents were married in my metaphor, I guess it's not technically illegitimate.

Other professions, such as law, medicine, electrical engineering, computer science, and mathematics, have distinctive pedigrees historically. Their pedigrees give them the status of being dignified and ordered professions.

Many professions emerging from the middle ages used spoken and written Latin. Latin was the language used by the educated, especially when communicating to persons of different tongues. The use of Latin had inherent advantages:

- Latin was a universally understood language of educated people.
- The use of Latin prevented common and uneducated people from seeing how the sausages were made.
- The use of and fluency in Latin became a badge of honor, proof of club membership.
- The uneducated could not corrupt the language, as they didn't speak in Latin.

For example, the language of lawyers is not of the common people. Lawyers originally spoke in Latin as a means of preserving their language from contamination by common folk, and this is still seen today with terms like *habeus corpus* and *quid pro quo*. The legal profession has an effective system in place to prevent outsiders from acting as lawyers. All lawyers must pass the bar examination in order to practice. And bear in mind that when non-lawyers represent themselves without legal counsel, they frequently lose their cases when the experienced lawyers cite technicalities.

The legal profession has been around a long time, whereas mechanical engineering is a relative newcomer in town. As a profession, mechanical engineering is fighting an uphill battle. The profession suffers today because of several major drawbacks:

- The language of mechanical engineering is imprecise.
- The language of mechanical engineering is not protected. Others—those outside of mechanical engineering—have ready access to it. Because outsiders use and abuse the language of mechanical engineering, it continues to be degraded.
- Mechanical engineering represents a broad area of practice. There are few fences and guard dogs to keep others from entering.
- The residents within mechanical engineering opt to cede

territory and to not repel invaders. Mechanical engineers don't play offense.

With this overview of the history of mechanical engineering in mind, I'll give you the pros and cons of the profession. Then you can decide if it's the career for you... or for your worst enemy!

Why You Should Become a Mechanical Engineer

Mechanical Engineering represents a unique and attractive career choice. In this epistle, I strive to emphasize just how special it is—few other professions come close—and I expound on why mechanical engineering should be seriously considered. Moreover, I will say it in three words: to develop a*rête*.

Arête is a Greek word that reflects the desire for wholeness in life. A whole person dislikes specialization and the localized concept of seeking efficiency via concentrating on one skill to the virtual exclusion of other skills. People who possess *arête* achieve a sense of global optimization. They enhance their ability to function in the world by being generalists. This one reason is so strong that it outweighs every other possible career choice or reasons against mechanical engineering. Mechanical engineering is the one and possibly only curriculum choice that wholly provides a student with an understanding of the physical and technical world and teaches how to function successfully within that world. The end result is becoming a whole person. In contrast, most other curricula options train a person to do tasks, tasks that are restricted to specific problem areas. For this one reason, if I had to decide on a major today, I would again choose mechanical engineering.

I suggest reading *Zen and the Art of Motorcycle Maintenance* by Pirsig [1]. He discusses the concept of *arête* and its relation to the achievement of excellence. Pirsig further states that "*Arête* implies a respect for the wholeness or oneness of life, and a consequent dislike of specialization. It implies a contempt for efficiency—or rather (a respect for) a much higher idea of efficiency, an efficiency which exists not in one department of life but in life itself."

Additionally, Lazarus Long, the protagonist in Robert A. Heinlein's novel *Time Enough for Love* [2], comments on the concept of *arête* this way: "Specialization is for insects." Long, a generalist, has the ability to live several thousand years while the rest of the misfits in the world, not pleased with his long life, wither and die. Long is forced to cope with a hostile world and must blend in to provide for himself while others try to hunt him down. A generalist is the only type of person who can live and be productive in a changing world. All the specialists resist change, are as parasitic as possible, and deny reality in order to benefit themselves and their corner of the hive.

The mechanical engineering curriculum provides an opportunity to study at the crossroads of physics, mathematics, materials, computers, fabrication, energy conversion, systems theory, component design, economics, tax law, human engineering, humanities, heat transfer, optimization, composition, graphics, instrumentation, chemistry, electronics, information technology, and philosophy.

For example, mechanical engineering students at the University of Illinois learn by using computers so much that they, historically speaking, consumed more mainframe computer time (measured in service units) per student than any other curriculum on campus, including computer science, electrical engineering, and computer engineering. Also, the mechanical engineering student receives rigorous training in mathematics, which is then integrated into engineering analysis and ultimately the physical world itself.

Mechanical engineers can handle mathematics. Another of Long's quotes comes to mind: "Anyone who cannot cope with mathematics is not fully human; at best, he is a tolerable subhuman who has learned to wear his shoes, bathe, and not make messes in the house" [2].

I do not wish to state that students in non-mechanical engineering curricula are denied the chance to become full human beings. Rather, I assert that the mechanical engineering curriculum training, exposure, and education provide an ideal opportunity to become a whole person.

Other justifications exist for the selection of mechanical engineering. One justification rests in the realities of living in a hostile world. The world is filled with insects. As a mechanical engineer, you will be on your best footing to prevail against insects. We live in a world populated with misinformed people—subhumans who will do their best to subdue you. Newscasters who sensationalize stories, pronouncing that wintry conditions will prevent your car's engine from starting. People who tell you that objects sink in water because they are heavier than water. Folks who believe the Earth is closer to the sun in summer and farther in winter and that's why summer is warmer. I could cite hundreds of other ridiculous proclamations, but in the interest of brevity I will move on. I will keep my bug spray handy, though.

Why You Should *Not* Become a Mechanical Engineer

Every profession has its snares, pitfalls, and dead-ends. I will discuss some that apply to mechanical engineering. My purpose isn't to deter you or scare you away, but rather to provide forewarning so that you can better avoid these downsides. Although I've never had the official job title of mechanical engineer, I was in academia for my entire career, and so for the purposes of this discussion I'll consider myself one of the group. We're really all in it together.

We Can't Get Any Respect

"No one can make you feel inferior without your consent."
~Eleanor Roosevelt

#1

Many people think they can perform mechanical tasks.

We live in a technological age where the typical person is expected to be able to perform mechanical tasks such as changing a car tire, lubricating a squeaky wagon wheel, and changing the air filter in a furnace. In our society, any person who lacks those basic skills is made to feel inadequate and undignified.

Thus, it is not unusual for most people to think they can accomplish what the mechanical engineer does as a profession. Whether they can or not is not the issue. The typical person says, “Oh, I could have done that.” They are convinced that they could have easily performed the mechanical task, and any effort to suggest otherwise is usually interpreted as a put-down.

A mechanical engineer’s life’s work can often be boiled down into one YouTube how-to video these days. This gives people further reason to believe they can tackle tasks that have previously been reserved only for those who know what they’re doing. The famous disclaimer, “Don’t try this at home,” has been replaced with ubiquitous “You can do it!” encouragements. Perhaps the career path of today’s mechanical engineer should include “YouTube star.”

#2

Persons without training in mechanical engineering feel qualified to meddle with and judge our designs.

When a mechanical solution is developed, the world feels compelled to adjust, tinker with, eliminate, substitute, and modify the various components of the mechanical design. Conversely, if the solution is composed of circuitry, machine code, economics, a medical prescription, a musical score, or an accountant's double-entry financial statement, as a general rule the world does no such thing.

Let me describe a consulting experience as an example. In the summer of 1981, I was retained by a company to perform an autopsy on a modestly complicated machine that failed to perform reliably.

I was called in to diagnose and fix, if possible, a faulty mechanical machine that moved wafers of silicon through an RF plasma chamber via mechanical vibratory techniques. In theory, the machine functioned according to vibratory feeder principles. An aluminum track, about five feet in length, was vibrated using an electromagnetic driver. The silicon wafers, about four inches in diameter, were intended to move across the beam. The purpose of the transport was to permit RF plasma etching in layers on the wafers. The process was then state-of-the-art. The concept was great except for one problem—the machine failed to move the wafers consistently. They would go only partway and then stop, or even back up at times.

A friend of mine was the Vice President in charge of sales. He couldn't sell machines that didn't work, so he called me, as he knew I was good at fixing things. My final diagnosis was that the problem was due to the original design being modified with many other designs.

The head of engineering, another VP in the company, was an electrical engineer (EE) who had many years of experience. Without telling anyone, he had specified the replacement of a two-polarity power supply of AC current with a one-polarity power supply. The one-polarity power supply was switched on/off in bang-bang fashion, yet the EE asserted that it didn't matter. In essence, the original electromagnetic excitement was sinusoidal. To save money, the EE substituted a cheaper square wave excitation. Unfortunately, a square wave has an endless number of higher harmonics, which distorted the vibrational modes of the track and caused the wafers to sporadically stop and back up. The change in electrical specifications was a prime reason the mechanical machine failed.

The purchasing department was also to blame as they saved money by changing weldment designs and vendors six times without communicating that information to the engineering department.

The production people in the plant were also at fault as they assembled and fine-tuned the machines on a non-level wooden floor. The factory was in an old warehouse with uneven floors supported by old wooden beams. When asked about the situation, the production people replied, "Oh, we don't bother leveling machines anymore. It doesn't seem to matter." They disregarded the fact that if the machine was shipped and reassembled on a different floor, the machine wouldn't always work. Commonly, companies purchasing the machines would ship to modern factories in Silicon Valley, factories with level concrete floors. The vibrational properties of the design varied greatly depending on the type of supporting floor and its being level.

Then the field service people put bolts in wherever and whenever they felt like it. As the device was based on vibratory feeder principles, it was as sensitive as a Steinway grand piano. You don't go drilling and bolting randomly on the soundboard of a grand piano and expect it to remain in tune. Just as all pianos in the production line will be different if they are made differently, so will all mechanical machines have their own sensitivities based on impromptu remedies.

Finally, the customer contributed to the faulty machine by doing anything and everything he could on an *ad hoc* basis to try to get the machine to run.

Of all who tinkered with the machine, the electrical engineer Vice President of the company was the most dangerous because he had minimal technical grasp while also having authority. His default behavior was to insist that the fault of the machine lay with a dirty vibratory track in the partial pressure plasma chamber.

It took me considerable time to troubleshoot, but I eventually learned that the dirty track was not the problem. The EE argued his dirty track hypothesis based on his tests. He would wipe down the track using alcohol and clean rags. Then the chamber was closed and depressurized. Amazingly, upon start-up, the wafers moved properly. Yet, after an hour or so, the wafers failed to move. The alcohol used to clean the track dissolved into the track's porous aluminum surface, releasing alcohol molecules for substantial durations in the partial pressure chamber (it was at 1/100 of an atmosphere) after the track was dry to the touch. These molecules formed a gas under the silicon wafer that caused the wafer to vibrate at a different phase angle with respect to the track's drive frequency. Thus, the track only worked for an hour or so following the alcohol application, until the gas dissipated. The VP was the one who advocated the use of alcohol to clean the track. Being in a hurry, the chamber was vacuum pumped to demonstrate the VP's dirty track hypothesis.

The machine was a basket case. Yet it was the mechanical engineer who designed it who took the heat. In 1981, these machines cost $180,000 each, and 80 of them were eventually scrapped. The company ceased to be traded on the New York Stock Exchange, and the outcome represented the death of an $8M per annum company.

As this anecdotal story illustrates, mechanical engineers are not always consulted regarding changes in our design specifications. The best mechanical engineering solution is rarely adopted as-is. It is simple human nature for the purchaser to chisel and cut corners. In the end, a cheaper and at times faulty solution is adopted. Yet the mechanical engineer gets full credit when the design fails.

#3

Many view mechanical engineers as superfluous, thus non-essential.

Because most people believe they are part-time mechanical engineers, the decision to hire a mechanical engineer is often made only when the person or corporation is too busy or doesn't want to be troubled.

For example, I now hire somebody to mow my lawn. The task is not beyond me. My reason is for convenience. This rationale applies to many tasks performed by mechanical engineers. The decision to hire a mechanical engineer is rarely based on the belief that a task is so difficult or specialized that only a mechanical engineer is qualified to handle the problem. When an employer fails to engage a mechanical engineer, there are no perceived consequences. In short, many believe that the job we perform is not important. Neither the task nor the mechanical engineer is considered to be vital.

When cutbacks must be made in an organization, the various professions within that organization are handled in different manners. Clerical personnel and even loading dock personnel can only be dismissed based on cause. (Somebody has to find and document a specific reason or mistake, which may be difficult to do.) Thus, layoffs are based almost entirely on seniority with all sorts of protection rules and safeguards acting to protect the employee.

Engineers, on the other hand, have virtually no mechanisms for protection from layoffs. Companies are free to lay off whomever they wish without citing a cause and without resorting to seniority protection. At a large Midwest agricultural tractor manufacturer where two-thirds of the engineers were laid off, the one-third left on the job were not the better engineers. They were the less effective ones who had the benefits of political connections and friends in high places.

#4

Our knowledge is not appreciated.

Mechanical engineers tend to be soft-spoken and are not adequate advocates for themselves. Take the raging debate on climate change and global warming. With little question, the unqualified have taken center stage in the debate. In contrast, those who have credentials in heat, thermodynamics, radiation, and systems theory are somehow perceived as unqualified to discuss these topics.

I devoted my education and professional life to obtaining an understanding of heat principles and related systems theory, all of which falls flat when I try to explain it to the public. Despite my credentials, any contribution of mine to the conversation has been unappreciated and not taken seriously. In my research concerning ice age causation, I lay out the scientific principles behind the cyclical warming and cooling of the earth. To me, the facts are undeniable. However, the controversy about global warming appears to be more politically-driven than fact-based.

Many outspoken pundits with zilch knowledge and expertise about heat transfer, radiation, thermodynamics, fluid dynamics, and systems theoretic principles presume they have answers regarding today's hot-button topics of climate change and global warming. I hold that few of these self-proclaimed experts could clearly explain the terms albedo, emissivity, Kelvin, latent heat, and Lyapunov Stability. By inference, such pseudo-experts dismiss my earned doctorate in mechanical engineering sciences and mathematics.

Bear in mind that mechanical engineering stands on three historic legs that are all rooted in the Industrial Revolution of the 1700s: the thermodynamics of heat, which encompasses thermal energy, heat transfer, fluid flow, and radiation; the mechanisms that permit heat to be converted into useful work; and feedback regulation principles, which allow people to control the steam engine's speed and power output.

Mechanical engineering embodies the thermodynamic principles that underpin Earth's heat and radiation dynamics. But alas, the self-proclaimed experts on climate routinely reject my ideas even before my first utterance.

In sum, the professional qualifications of mechanical engineers are severely discounted in our culture. Most people are convinced they know more about climate change than me. They judge my work on ice age causation and climate before I've even gotten through my explanation of the ice age. And I promise my explanation isn't anywhere near as long as the actual ice age! It just seems that many people have an instinctive negative reaction to anyone who is about to espouse actual facts.

My research on climate change will continue to be contested and questioned. I can accept that. After all, the scientific method is rooted in skepticism and questioning. What I have a harder time accepting is that my credibility is dismissed even before I make my argument. My conclusion is simple: There is little point in amassing credentials only to be discredited at the onset.

#5

We're not magicians.

Mechanical engineers don't have a sufficient supply of magic tools or up-to-date buzz words. Computers, CAT scans, white robes, double-entry books, fiduciary contingencies, and blinking lights aren't part of the package. The general public is not impressed by words like heat, air conditioner, engine, spring, oil, work, horsepower, and calorie.

The "tricks of the trade" that we come up with are not enough to convince the world that we are going to solve their problems. Electrical engineers and computer science professionals have the computer with all its mystical powers. The mechanical engineer is often excluded from promotion to high places because the job usually goes to someone with "high-tech" credentials and tools.

In our political and dark age world, the nod is usually given to the man or woman with the magic tricks. Mechanical engineers don't have much public image of magic left. Most of our tricks were given away, usurped, or revealed decades or centuries ago. We're not considered to be part of the high-tech generation.

I began this reflection in the mid-1980s. Upon returning to this manuscript in 2019, I was surprised by how accurate my forecasts were more than 30 years ago. The situation for mechanical engineers has not improved but instead has worsened. The advent of 3D printing is revolutionizing manufacturing and has enabled non-mechanical engineers to do prototype development. The perceived need for the mechanical engineer becomes less critical every day. Perhaps if we wore wizard hats and waved around wands, we'd finally get some attention.

#6

We earn less money than we are worth.

The career advancement of mechanical engineers is slow compared to the high-tech engineering and computer industries. As a consequence, with a smaller proportion of mechanical engineers at the top rungs of the ladder, a mechanical engineer at a lower rung is less likely to find an advocate at the top willing and able to pull for him/her. The top is occupied with those who tend to pull for their own kind when given the choice.

As a group, we sell ourselves out too cheaply. Entry-level

salaries may be comparable to that of many other professions, but it is not uncommon that the mechanical engineer is stuck on a flat salary track that hardly keeps pace with inflation. After a decade on the job, the mechanical engineer is often paid at the same rate or even less than entry-level people. This is known as salary compression.

When asked about his consulting fee regarding the collapse of the Tacoma Narrows Bridge in 1940, Theodore von Karman said, "I started to say fifty dollars a day—my standard fee for government work at that time—my colleagues quickly hushed me up. They said they would do the talking." [3, page 215]. Von Karman's colleagues were civil engineers. They bargained for a sizeable percentage of the value of the bridge, which was insured for six million dollars, and they got it. I realized that architects and civil engineers who build bridges and other giant structures approach their compensation in the right way, while we mechanical engineers are only elevated laborers insofar as fees are concerned.

Mechanical engineers are often paid a flat hourly or daily fee whereas many other professions receive a percentage of the action, which can be quite large. In the case of lawyers, the action in lawsuits can be hundreds of thousands or even millions of dollars. Attorneys receive an average of 30% in addition to the client's payments in advance for all out-of-pocket expenses such as court recorders, hiring of expert witnesses, and filing fees. Real estate agents handling large commercial sales often get 5-10% of the entire price of the building just for creating the sale.

Lawyers and doctors have been successful in perpetrating the myth that no two problems are the same. Instead, every client and patient is told that the lawyer or doctor must see him, as nobody lesser than that will do. It is not unreasonable to estimate that doctors charge $100+ for a 15-minute office visit, and doctors don't forget to bill that patient. Also, the public eagerly sits in line to be seen so that four or so patients per hour is not unusual. The reality is that people in these professions are raking in upwards of $400 per hour! And to think, we haven't even started on the big hitters like the open-heart surgeons. I doubt if there is more than a handful of mechanical engineers in the world who make $400 per hour. Doctors on the other hand routinely make this much and more. My dentist even has the nerve to put out a sign saying that all services are performed on a cash basis due to the high cost of billing.

#7

Patents are commonly pathways to poverty, especially for the sole mechanical engineer.

Eli Whitney (1765-1825), the inventor of the cotton gin, died a pauper. His patent was widely infringed. Whitney was forced to seek remedy in numerous courts, and his legal fees drove him into poverty. While patents may be appropriate for some with deep pockets, the little guy on his own should exhibit caution.

I am reminded of the story involving the inventor of the quick release button on ratchets. The inventor was an employee of Craftsman Tools, then a branch of the now defunct Sears & Roebuck store chain. He submitted an internal memo to his employer. His idea was rejected. Sears/Craftsman then took the idea and developed the quick release ratchet wrench. The man later sued and won. So I suppose one suggestion for success is to let somebody steal your idea. After they perfect and market it, then you sue them.

As an aside, I strongly advocate keeping a design and engineering journal to protect yourself and your ideas. Make your suggestions in writing. Keep copies of your correspondence. And make copies of those copies, then lock them in a safe.

#8

Our turf is often encroached upon.

Those without credentials tread upon the mechanical engineering domain all the time. The laws to protect the mechanical engineering domain are a joke. If a newscaster attempted to explain legal issues, the legal profession and state bar association would have the newscaster and station run out of town. Doctors would likewise have a similar retaliation. In contrast, newscasters are allowed to make absurd claims related to mechanical engineering topics.

If a mechanical engineer solves a difficult problem, his or her answer will usually be treated with some degree of skepticism compared to the work of other professions. Yet who questions, for example, a CPA and his or her conclusions? Who questions journeyman welders? Who questions a victorious military general? Who questions a chimney sweep? Who questions a hair stylist? But there are many different ways in which our credibility is doubted.

For example, mechanical engineers say that radial tires and non-radial tires should not be mixed on a vehicle. Do you think the public believes us? A considerable percentage of people think that tires are tires and that this suggestion is really a trick to force them to have to buy a whole set of tires when they only need a pair.

Another example is that automobile antifreeze is supposed to be mixed or diluted with water. Do you really think the public accepts the fact that a mixture of water and antifreeze exhibits both a higher boiling point and a lower freezing point than a coolant system filled with straight antifreeze? In fact, it is not uncommon for consumers to use straight antifreeze in the belief that if a little bit is good, a lot more is better.

A third example brings us back to a favorite topic of mine: the attention-getting wind chill factor. I once heard a newscaster say that the wind chill was going to make it harder for automobiles parked outside overnight to start. People who are not heat transfer experts should not pass themselves off as such—especially in such a public forum—because it only increases the spread of misinformation. Wind chill factor is a measure of the time rate of heat loss from a heated or warm object, such as exposed human skin. Wind chill factor has no bearing on cold objects, such as cars parked outside for long durations. Hence, the ability of an automobile to start on a wintry day is in no way affected by a low wind chill factor but merely by the ambient temperature of the outside wind (or air) as indicated by an ordinary dry thermometer. Do you think most people really believe such a suggestion? The general public associates coldness with how they feel, so this pronouncement by mechanical engineers is not taken seriously. People believe their senses much more than mechanical engineers (who insist that "radiators" be more accurately called "convectors," thus further eroding our credibility).

I heard a meteorologist on a TV newscast declare that a current year's winter was 8% warmer than that of a previous winter. In my view, the meteorologist believed this was "proof" of the global warming theory. I don't doubt that the public accepted the weather person's statement. But if a mechanical engineer had pointed out that temperatures must first be converted to absolute temperature scales (such as Kelvin or Rankin) before the concept of percentage of cold or hot even makes sense, then the public would shake their heads at how petty and pedantic we are. What the meteorologist should have said is that the current year's average winter temperature was 8% higher than the previous winter. Seems like a minor difference (some might even say petty and pedantic!) but it is at least factual.

While weather predictions are a bit of a joke, people still eagerly tune into weather stations. They'll settle for any scrap of information, no matter how far from the truth. Yet for some reason, the actual truth coming from the mouth of a mechanical engineer isn't met with the same enthusiasm. Perhaps because we are dry and complain too much?

What's more, our academic territory is being invaded with looters who stake claim to the better parts of mechanical engineering. Pushback from the ME community is all but nonexistent.

Control systems started in mechanical engineering, but it is now the virtual property of electrical engineers. Kinematics now has to share the limelight with the computer. Theoretical and Applied Mechanics has laid claim to finite element analysis, materials, and vibration theory. Aeronautical engineering shares applications of fluid flow. The area of Operations Research has been invaded by mathematicians, computer scientists, civil engineers, and specialized systems engineering groups. Mechanical engineering has no exclusive claim to Operations Research whatsoever, and the claim of industrial engineering is in dispute.

In response to all of this, the national trend of mechanical engineers is to shrink into a small circle of wagons around the thermodynamics campground. The high priests of thermodynamics are in charge of mechanical engineers, and they seem oblivious to the real and permanent losses that mechanical engineering is suffering.

It is my speculation that the above statements will be met with hostility by the mechanical engineering community. I'm ready for it!

Furthermore, I challenge anyone to find any *ASME Transactions* paper written in the history of the ASME that has been cited more times than the 1961 ASME paper by R. E. Kalman and R. S. Bucy [4]. I contend that more than 80% of mechanical engineering faculty (at a random collection of prominent schools with mechanical engineering programs) are unable to comment intelligently on said paper. I also allege that more than 90% of graduating BSME seniors cannot comment intelligently on the Kalman-Bucy work. The point is that the Kalman-Bucy paper and its results are the most significant finding ever published in *ASME Transactions*, yet it is virtually unknown to mechanical engineers because it is out of the territory of the thermodynamicists. Hence, thermodynamicists are blind to the world about them.

It is also interesting to note that electrical engineers would not fail the above proposed tests for mechanical engineers. Some argue that Kalman was an electrical engineer. The logic behind that assertion escapes me. Kalman earned three degrees in mechanical engineering, obtaining his doctorate from Columbia University. The acclaimed Kalman-Bucy 1961 paper was published in *ASME Transactions* of all places. What else does one have to do to be a mechanical engineer?

An additional note regarding loss of territory: The area of CAD-CAM (computer-aided design, computer-aided manufacturing) and computer graphics design is currently being contested by computer scientists as they try to wrestle it away from the fragile grip of mechanical engineers.

#9

We perform *a la carte* services.

All engineering solutions are subject to economic justification. The major distinction, however, is that for most engineers and designers other than mechanical engineers, the entire solution or design is accepted or rejected as a package. As examples, this all-or-nothing approach is characteristic of chemical engineering, mining engineering, petroleum engineering, aeronautical engineering, and nuclear engineering. All competing design packages may be equally ineffective, so to speak. The engineer who deals in the realm of invisible things has a shot at selling his or her design solution as a single entity.

The visible engineers (mechanical engineers who design

visible products) face the situation where their entire solution is a candidate for being picked apart. In this case, the client or purchaser feels justified in picking, choosing, and blending.

The French term *a la carte* comes to mind. It is human nature to take the good for granted and to reject the bad portions. The mechanical engineer and his/her solutions are forever being picked over and abused. By diluting or removing potentially integral parts of the solution, the solution is degraded. Or *dégradé* as the French would say.

#10

People think global warming is our fault.

Fossil fuels are the lifeblood and foundation of our society.

The essence of mechanical engineering is the creation of machines that convert heat into useful work, and a second emphasis is the creation of machines that transmit useful work. Climate change pundits have set their sights on the elimination of our societal dependence on fossil fuels. While these fossil fuels opponents are unlikely to prevail, they will still aim to make life uncomfortable for those in the fossil fuels business—and notably for mechanical engineers.

#11

We seldom get promoted up the management ladder.

Mechanical engineers are team players in most industries. Those who manage engineers tend to have business administration and/or finance backgrounds. Management types are expected to advocate new and creative ways to approach business problems. Management types who fail to be innovative and even adversarial are viewed as ineffective and thus unnecessary. Mechanical engineers, being consensus people, tend to be passed over for upper-level positions.

But Maybe We Don't Deserve Any

"Part of me suspects that I'm a loser, and the other part of me thinks I'm God almighty."
~ John Lennon

#12

We feel compelled to explain our trade secrets to outsiders.

Since many mechanical engineers sense that non-mechanical engineers think they are not vital, we commonly feel we must explain "why." In explaining our reasoning, we cause these non-mechanical engineers to think they understand, which then renders us unnecessary. The feeling that the mechanical engineer is dispensable arises both before the fact and after the fact, hence *a priori* and *a posteriori*.

The medical profession, as an example, deliberately hides any remedy or design solution. The most common evidence of this is that medical prescriptions are written in Latin. Also, it is seldom in clear Latin, but rather a hodgepodge of Latin, abbreviated words, mathematical symbols, and deliberately unintelligible handwriting. Medical doctors, it seems, have been trained to mask their remedies. It is obvious that their teachers have taught them well. The remedy is not clear to the average lay reader or patient, and certainly the reasoning behind the why is much less clear than that, perhaps even nonexistent.

Computer science types don't explain the inner workings of their machines or code to the general buying public. Clergy, at times, speak in Latin. The legal profession makes a handsome living by refusing to explain to lay people the priorities of how various laws are applied, yet a good lawyer certainly knows which laws come first and which come last.

Mechanical engineers, on the other hand, often feel compelled to reveal all the details of the design solution, complete with a clear explanation as to the why. We go to great lengths to sort things out and to explain principles and reasoning. We not only show you the inside of the sausage factory, we show you how we designed every single machine in it.

#13

We sell ourselves short.

Mechanical engineers, as a group and a profession, have neglected to originate, disseminate, and exploit rumors that this profession is perilous and that its practitioners are worthy of special honors. Some groups who have been successful at this are milling machine department inspectors, economists, lawyers, doctors, accountants, high steel workers (who do aerial rigging), rodeo bull riders, deep sea divers, crop duster pilots, electrical engineers, and bomb disposal demolition experts.

In addition to mechanical engineers, a few other isolated groups have also failed to exploit suitable myths to keep the common people at bay. For example, public grade school teachers. I happen to be married to a career kindergarten teacher. When asked about whether teaching five-year-olds is difficult, my wife modestly replies, “Oh, it’s easy. I enjoy it.” A preponderance of parents is absolutely convinced they know how to teach kids because they have been to school once themselves and know all about it. If anyone doubts this, just ask some parents.

#14

We are all too easily replaced by a new crop of college graduates.

Mechanical engineers don't hang out shingles. Instead, we are typically hired (and fired) by companies who must be practical and business-like. Individuals requiring an expert, such as those on trial for a crime, may be fooled by a fancy reputation and will pay accordingly, but companies who have to stay in business don't hire the "fastest gun in the West" at outrageous prices. Instead, they tend to hire entry-level people from the masses of fresh college graduates and let the chips fall where they may.

These entry-level employees are worked hard. The winners get a modest reward. Those who fall short are reassigned to lower duties or are discharged.

Once the winners burn out from this treatment, it is not uncommon for the corporation to discard them too. We live in a society with a throwaway mentality. Unfortunately, it is sometimes people who get thrown away when they are no longer valued. After all, a new crop of entry-level recruits is readily available. One admonition I have come to believe in is "Talent is cheap." Ample numbers of bright minds are willing to work for others at modest salary levels.

#15

There is no means to honor the best mechanical engineer.

There is no effective forum in the profession to prove that a given mechanical engineer is the best out of all the mechanical engineers. Other professions have been successful in convincing the public that somebody is the best in their respective field. It is not uncommon to hear phrases like, "the best trial lawyer," "the best architect," "the best open-heart surgeon," "the best accountant," "the best tenor," etc. When was the last time you ever heard someone imply that he or she had the hardest and most vital mechanical engineering problem and that he or she was going out to hire the best mechanical engineer to take on that problem? In the other professions, the establishment of a "best" allows that person to charge accordingly. This permits phonies to claim to be 85% of the best and to accordingly charge 85% of the best person's billing rate.

If a person is in fact able to name one, the best mechanical engineer is usually labeled as a star in spite of mechanical engineer training, not because of it. As an example, Lee Iacocca of Ford and later Chrysler was a mechanical engineer. Iacocca started in mechanical engineering, switched to industrial engineering (IE) for his BS (Lehigh) and then obtained an MS (Princeton) through a thesis with mechanical engineer flavor concerning a hydraulic pump, which is close enough to be considered as a mechanical engineer. He made it in the automotive world by being the creator of the Ford Mustang. It was a marketing and styling feat that had very little to do with his mechanical engineer skills. The chance of you or I duplicating his feat is quite miniscule. In Iacocca's autobiography [5], he gives some insight into his success: "The day I'd arrived [at Ford for his first day on the job as an engineer], they had me designing a clutch spring. It had taken me an entire day to make a detailed drawing of it, and I said to myself: 'What on earth am I doing? Is this how I want to be spending the rest of my life?'". He immediately left engineering and entered the sales side of the business. One day as a mechanical engineer was enough for Lee Iacocca.

Furthermore, no mechanical engineer has ever received a Nobel Prize. Due to the current structure of Nobel Prize categories, no engineer will ever be considered. The categories are set up according to traditional categories: economics, literature, medicine, physics, and peace.

Rudy Kalman (1930-2016) earned his PhD in mechanical engineering from Columbia University. Kalman solved a problem that was originally posed by Norbert Wiener in 1932. Kalman's solution of the Weiner filtering problem was mathematically elegant to the point that the solution was done in one paragraph in a footnote of a paper on another topic [6]. Kalman's solution had profound implications for society and the advancement of technology. For example, the Kalman-Bucy filter [4] enabled the first soft-landing on the moon, which took place in 1963. To the best of my knowledge, Rudy Kalman never received recognition like Norbert Weiner, who only asked the question.

Many conclusions can be drawn from this anecdote, but two will be noted: It is the nature of mechanical engineers and the creative process that the most elegant solutions are, in fact, the simplest in concept; and mechanical engineers are systematically excluded from consideration for honors such as Nobel Prizes.

Is a Nobel Prize too much to ask for?

#16

We failed for a century to explain how and why the bicycle works.

The truly great mechanical engineering inventions, such as the bicycle, were developed by clever and yet common people. Even though the modern bicycle has been around for well over a century, it was commonly believed that the bicycle's inherent stability and handling ease were from gyroscopic torques due to the rotation of the wheels. Dr. David Jones, an English chemist, built a non-gyroscopic bicycle circa 1970 as a lark [7]. He negated the gyroscopic torques by mounting a second wheel next to the regular front wheel. The new wheel was mounted slightly higher on the side of the front fork, so it did not touch the road surface. This extra wheel could be spun backwards, more or less cancelling the gyroscopic effects of the normally forward spinning front wheel.

Jones' simple experiments demonstrated that gyroscopic

torques had negligible effects on the bicycle's operating characteristics and general ease of handling. Hence, the bicycle became an open mystery. It is true that computer simulations existed in the literature, notably the work of R. Douglas Roland done at Cal Span Research Laboratory [8]. The difficulty was that computer simulations may duplicate behavior, but they don't explain it. In the late 1970s, approximately 20 peer-reviewed articles attempted to explain how and why a bicycle worked. Hypotheses were put forth, but no consensus existed. Most of the papers were in disagreement. However, reality suggested that the literature was based on faulty reasoning. My paper with Astrom et al [9] provides an overview.

The confusion over the bicycle question suggested a weakness of the mechanical engineering profession. How can you take any group of professionals seriously who claim to have expertise in mechanical matters yet are incapable of stating why and how a bicycle functions, especially when the bicycle is mechanical, visible, and has been around for more than 100 years?

The bicycle has no secrets (or so the public thinks). Every force, mass, torque, inertia, input, and road interface action is visible to the eye (or so the public thinks). The reasoning can be explained with sophomore-level physics (or so the public thinks). Given how simple and plain a bicycle is, how could anyone trust mechanical engineers with something more complicated?

#17

We hold our winter annual meetings in undesirable places.

The mechanical engineering professional society, the American Society of Mechanical Engineers (ASME), traditionally holds its annual winter meetings in garden spot cities such as Philadelphia, Detroit, Pittsburgh, and New York. Other professional societies for electrical engineers, doctors, orthodontists, accountants, etc., select locations such as Las

Vegas, Honolulu, San Diego, and Tampa-St. Petersburg for their winter meetings.

To a mechanical engineer, winter means snow and slush in Pittsburgh. To other groups, winter means going south to palm trees and white beaches with the IRS sharing a good portion of the tab. For those groups who actually choose snow, they do it correctly by scheduling meetings in Aspen and Steamboat Springs. Still, most meetings are scheduled in beautiful sunny cities like Sanibel Island, Acapulco, and San Juan.

I once attended a winter meeting in Cleveland, Ohio. After dinner, I asked the hotel clerk what attractions were nearby for fun and entertainment. The desk clerk's advice was to stay within the confines of the hotel after dark due to the high risk of being mugged on the nearby streets. The hotel bar didn't even have fruity drinks with umbrellas.

#18

We are typically paid by a company rather than a client.

Mechanical engineers are almost exclusively paid salary reported on a Form W-2 for income tax reporting purposes. In contrast, persons in other professions who hang out a shingle are paid income reported on Form 1099-MISC.

In terms of income tax reporting, the differences are striking. Persons in self-employ can form sole proprietorships, LLC's, and S-Corporations. Those who receive W-2 statements have limited means to write off business expenses.

As an employee we are working for "the man" instead of ourselves.

#19

We don't handle failure well.

In the early part of World War II, a Japanese vessel sailed home from the Sea of Japan with 20 or so unexploded torpedoes sticking out of its hull like cigars [10].

How could something like this happen?

Well, here's the story. A U.S. submarine had surfaced in the Sea of Japan and, to its surprise, saw a disabled Japanese freighter sitting like a duck in calm water. The U.S. submarine had all day to line up broadside at close range. The submarine took the time to calibrate sightings, check the activation mechanisms and detonators, and have multiple officers witness each task. They used all but two of their torpedoes.

The submarine fired more than 20 torpedoes at the freighter, but none of the torpedoes detonated. Instead, they stuck in the freighter's hull, plugging the holes they had made. The U.S. submarine captain decided not to fire the last two torpedoes, preferring to take them back to Pearl Harbor for scrutiny. Upon the return of the submarine, the captain was threatened with court martial for his decision to not fire all his live rounds at an enemy ship.

Now the Japanese knew more about our torpedoes than the United States did because we had given them several dozen. The U.S. valued its torpedoes so highly that none were taken apart but were only used in combat. Hence the duds were usually interpreted as misses when many actually hit and penetrated the target but didn't explode.

The problem turned out to be a faulty mechanical design of the detonator, and secrecy prevented the detonator design from being examined by anyone other than members of a closed group who all repeated the same error in their mechanical envisioning of the detonator design.

This is a classic example of how a supposed mechanical design error was denied and covered up for a long time. It was obvious to most officers and crews of the U.S. submarine fleet that many American torpedoes were duds. The officials, on the other hand, claimed that nothing was wrong with the torpedoes and that the problem was bad aim on the part of the commanders.

#20

You will have to look long and far to find a statue, bust, or building named in honor of a mechanical engineer.

I can speak about the University of Illinois, as I spent three decades there on the faculty. Over the past century and a half, many of the campus buildings have been named after leaders in respective fields. The Electrical Engineering Building was renamed to honor Dean William Everett. The Physics Building was renamed Loomis Laboratory. Talbot's name was put on the building that housed Theoretical and Applied Mechanics. The Civil Engineering Building was named in honor of Nathanial Newmark. The building that housed the Department of Mathematics was named Altgeld Hall. The list could go on.

I am reminded of a story told to me about William L. "Bill" Everett. He was Dean of Engineering when I was hired in 1968. Prior to becoming Dean of Engineering, Everett served as Head of the Department of Electrical Engineering at Illinois. When asked to define electrical engineering, his answer was simple and direct: "Electrical engineering is whatever my faculty is doing today."

Dean Everett did not place fences or barriers around electrical engineering. Instead, his faculty had the green light to expand and go wherever they wished. Moreover, the definition of electrical engineering was dynamic and ever-changing. I am struck by the word "today." I submit that if a similar question had been asked of the Head of Mechanical Engineering, the answer would be quite different. The answer would have been both restrictive and detailed.

But alas, the names of both the Mechanical Engineering Building and the adjacent Mechanical Engineering Laboratory have remained unchanged. The culture of mechanical engineering can't even imagine honoring a former faculty member or a student who once walked the hallways. Mechanical engineers have only renamed the department itself, now the

Department of Mechanical Sciences and Engineering. They prefer to obscure their name, as opposed to presenting the mechanical engineering profession in a place of honor. They are retreating from their heritage, as opposed to being proud of it.

#21

We tend to invent things that elevate others, thus diminishing ourselves.

A century ago, horses were still in common use. Handling and controlling horses was not for the timid and frail. Then the horseless carriage came along. Early automobiles, just as with horses, required strength and mechanical aptitude. But then mechanical engineers invented three things that changed the world:

- the electric starter
- the automatic transmission
- power assist for steering and braking

Now multitudes of people can operate cars and feel fully human. Conversely, as the population has been elevated through mechanical engineering efforts, mechanical engineers themselves have seen their advantages erode. Mechanical engineering is a profession that brings about its own demise.

#22

We are social misfits.

Mechanical engineers have, comparatively speaking, a hard time making and keeping friends on a social basis. The reason for this stems from the fact that we want to explain mechanical things to others. In doing this, we mistakenly believe that those around us actually desire to know about the matter. We're so deep into fluid dynamics and vortex shedding that somehow we miss our audience's eyes glazing over. The actual truth is that the majority of the public can't comprehend mechanical things or concepts, nor do they care.

Though others may hide their inadequacy well at times, often by pretending to be proficient at mechanical things, the last thing they want is to be lectured to. They really don't want to know, and they prefer not to be so vividly reminded of their shortcomings.

I'll always remember an incident that occurred in the midst of a chat about canoeing. In response to a question, a lady claimed that she owned a flat bottom canoe and that it differed from a white-water canoe only in the matter of the keel design and shape. She proceeded to say that the shorter keel of a white-water canoe permitted easier turning, whereas the lake canoe yielded a greater tendency for maintaining a straight course.

This particular lady was misinformed. A primary distinction between lake canoes and white water canoes has to do with the cross-sectional profile. Lake canoes by design have flat bottoms. In contrast, white-water canoes have rounded bottoms.

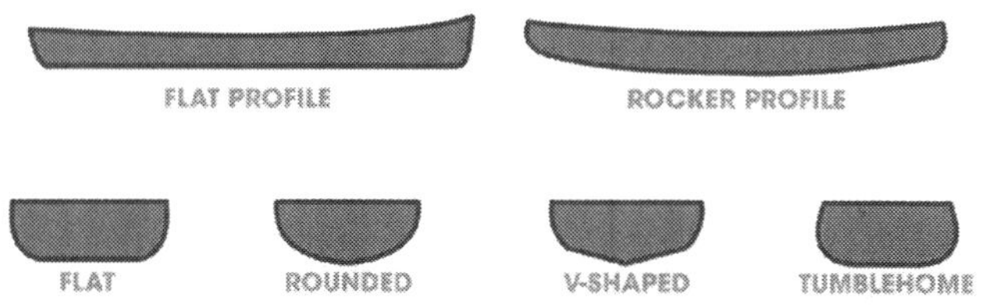

The sketches above show, respectively, flat and rounded canoe profiles. The flat bottom or lake canoe is more stable in smooth or lake water conditions. The rounded canoe profile requires greater canoeing skill, but it is also less prone to being overturned in wavy conditions. A wave striking a rounded object, such as a floating log, will not tip or overturn the log. In contrast, a flat bottomed object is more likely to be overturned in rough water.

I proceeded to explain how *metacenters* are calculated as the vertical projection of the centroid of displaced water while the vessel undergoes a small angular perturbation as it (the vertical projection) intersects the vessel's rotated axis of symmetry, how a more rounded cross section yielded a shorter *meta-centric height*, and that the subsequent ability of the white-water canoe is less responsive to the dynamic action of waves. Quite frankly, this lady was not pleased to find out she was in vastly over her head. Moreover, she resented the messenger who brought the news of her inadequacy. The mood of the chat was not lifted by my clear and informed discussion but rather it became heavy and depressing to say the least. She was clearly not impressed with my knowledge and professorial skills as I had expected; rather, she seemed to consider me a pompous ass. This came as an enormous surprise to me.

#23

Mechanical engineering, as a profession, is still fighting battles that ended a century ago.

The national organization ASME was founded to protect the public from unsafe engineering practices. Pressure vessels, such as those related to the steam engine, required standards and oversight to assure safety. The world has changed since the 1800s. Consumer products come from all over the world. Few products are designed by licensed mechanical engineers, those with the designation Professional Engineer (PE) behind their name.

Over a century ago, ASME adopted a Code of Ethics. That Code of Ethics has been rendered irrelevant. All sorts of people, both qualified and not, design products and provide the equivalent of engineering services. The name or label might be different, but the work of mechanical engineers is commonly done by others lacking training and qualifications. Products are marketed with an eye towards lowest price, not quality.

Other fields are experiencing astounding growth—and without regulation. I point to the area of Information Technology. The growth and new developments are coming so rapidly that half-lives are measured in years and possibly months. In contrast, progress in mechanical engineering is miniscule. The ASME old guard is worrying about ethics, and that battle ceased over a century ago. Lawyers, courts, marketing, and off-shore competition set the standard for what is designed and sold to the public.

We Are Victims

"Great spirits have always encountered violent opposition from mediocre minds."
~Albert Einstein

#24

We lack a forum in which to assert ourselves.

We don't work in a public adversarial forum, such as that enjoyed by the legal profession. It is the adversarial forum that allows lawyers to get the fastest-gun-in-the-West reputation. Mechanical engineers, on the other hand, work in an altruistic forum where ultimate truth is sought. Any mechanical engineer who adopts adversarial tactics, such as being critical of an opponent's view, is considered—by both the profession and society—as being unethical and self-centered.

For almost a century, the mechanical engineering profession has been "protected" by ethics. This prevents public disputes on the supposed grounds that it would tarnish the mechanical engineering image. Somehow the legal profession has worked around that and has done quite well, although it's arguable that the legal image is tarnished. We are not encouraged to assert ourselves and therefore we usually don't.

#25

The world remembers the errors of mechanical engineers.

Mechanical engineers are required to be accountable to reality. Politicians, lawyers, economists, electrical engineers, and computer science types, however, are so skilled and deceptive as to create a false standard. Theirs is a game where they avoid or postpone indefinitely being brought to answer for their errors. If a computer malfunctions, that proves you need a bigger computer and more electrical engineering and computer science professionals. Yet if a mechanical component fails, the public tends to think the mechanical part should be replaced entirely with a computer-commanded device.

It's a fact of life and a part of human nature that people are remembered for their errors, not their successes. Sir Winston Churchill noted that everybody told him what was wrong with his brick wall. (In the 1930s, Churchill devoted considerable time to the building of a massive brick wall in his yard [11].) People visiting his home pointed out every minute error. The bricks were not level, the bricks were the wrong color, the mortar was not uniform. Not a single critic bothered to tell Churchill that it indeed was a brick wall as planned, and that it would function successfully as a brick wall for as long as its builder intended.

Winston Churchill working on his brick wall.

It is funny how the only people with time to critique others' brick walls are those who haven't built any brick walls of their own.

#26

Mechanical engineers are generally culpable.

When a mechanical design fails, the finger gets pointed at the mechanical designer. In contrast, many other professions have avoided this.

If things go wrong, many professions commonly exploit that as the hazardous duty myth and use it as justification for longer robes, more research funding, and fewer questions. Consider the case of career politicians, who frequently break their campaign promises but continue to get re-elected, managing to pin the blame for the failure on the opposing political party.

Yet if a device designed by a mechanical engineer fails, the usual conclusion is that the mechanical engineer was incompetent.

#27

The work and mistakes of mechanical engineers are difficult to hide.

Many other professions—notably doctors—protect their colleagues despite gross mistakes. Autopsies and hearings for dead patients do exist, but they are strictly closed to the public. Medical autopsies are private out of respect for the dead, as are all laboratory reports. I am reminded of the old joke: What do you call the person who graduated lowest in his medical school class? The answer is "Doctor." Nobody ever flunks out of medical school. Once a person is admitted, if they fail a course, they keep repeating it until they finally pass. Even the worst of medical students end up being called doctors.

Bad doctors also don't get fired. The public seldom even gets to know who is a good doctor and who is a bad doctor. The insiders all know but rarely blow the whistle on a bad doctor.

An old saying is appropriate: "Doctors get to bury their mistakes."

Electrical engineers have a similar advantage. Due to miniaturization and the high price of copper scrap metal, the monuments to the stupidity of electrical engineers are often quickly and silently removed from the public's eye. The public seldom sees a computer scrap yard; however, they do see scrap yards filled with junked mechanical machinery such as conveyer belts, machines, punch presses, toasters, appliances, and automobiles.

When mechanical things wear out, fail, or become obsolete, those objects tend to be discarded, but the cost of removal often requires that the objects stay in place for all to see. Thus, the products mechanical engineers have designed are not always given a decent burial when they reach the end of their useful life.

Mechanical engineering products are too visible and too publicly available. As a consequence, once we solve a problem, our solution is open for all to second guess. The druggist, doctor, nuclear physicist, day care provider, church organist, etc., all work in such a way that the public is not privy to the details of the design, product, or service. Poor practitioners of these professions may be able to avoid blame for their mistakes. But bad mechanical engineers often get fired, or possibly worse, become lepers.

Let me explain. A former student remarked about the mechanical engineer identified as the cause of a product recall. If a poorly or defectively designed product became subject to a recall, the responsible designer would be banned and shunned. If not fired, the responsible designer was never again trusted. The responsible designer of a recalled product was treated as a leper. The leper stigma was permanent.

Mechanical engineers lack the benefit of a hearing to fully assign proper blame. A defense is not allowed. The conviction is certain based on the evidence—the design was recalled.

#28

Mechanical engineering designs are not easily protected. Reverse engineering is commonplace.

Others can gain insight from the mechanical engineer's work and apply this insight to their own product design. The knock-off or clone becomes competitive with the original. This process is commonly referred to as reverse engineering. Because of visibility, mechanical designs are especially prone to reverse engineering. This, of course, puts the copycats in a situation where they can undersell the original because they have a lower breakeven point (as the copycat doesn't pay for the original engineering development or contribution).

When a mechanical engineer's design is infringed on by a competitor, we have little recourse. Such designs are inherently visual, and visual designs, when patented, are difficult and costly to defend. In order for any patent to be granted, three conditions must be met: novelty, non-obviousness, and usefulness. Mechanical engineering designs are often considered obvious and thus not protected by patent.

To compound matters, many copycat infringers cut corners and market substandard products. When a substandard clone product fails, the creator of the original design becomes tarnished by association. An example of this is the hoverboard, which was invented and patented by Shane Chen in 2012 [12]. Soon after his invention became public, factories in China started to manufacture cheap knock-offs in violation of Chen's patent. The products flew off the shelves, but ended up sending thousands of kids to the E.R. [13]. The weaker motors and low-quality batteries left the devices underpowered and unstable, making riders more likely to fall. The counterfeits were also more likely to catch fire. Unfortunately, Mr. Chen's sales never caught fire.

#29

We are purposefully excluded from jury duty.

Lawyers have arranged a system suggesting that their work (in the case of trial work) is ultimately decided upon by a judge or jury. Anyone with anything on the ball is systematically excluded (rejected, even without cause or stated prejudice) in the jury selection process, as each side has a large number of available options to reject prospective jurors. This strange situation occurs because one side or the other is always weak, and whichever side is weak, or perceives itself to be weak, doesn't want intelligent and trained problem-solvers on the jury deciding the outcome of the trial. Consequently, technical, analytical, and broad-minded people like mechanical engineers are almost always dismissed from jury selection. Lawyers on the side with a weak case will reject the mechanical engineer, not desiring an analytical problem-solver who can sort things out and be able to explain every precise detail and logic to fellow jurors behind the closed jury room door.

One might conclude that favorite jury duty candidates are those who wear bow ties, swear by electric lawn mowers, form the basis of their world opinion upon what they hear at the beauty parlor, and enjoy the Lawrence Welk Show. Analytical thinkers and problem-solvers need not apply. The result is that the work of a lawyer is judged by rank amateurs who have limited prior experience, scant decision-making skills, and a stranger relationship with each other.

The mechanical engineer, in contrast, scrutinizes and arrives at decisions using a demanding, rational process. Although to be fair, this argument applies to other technical professions as well.

#30

Mechanical engineers are not recognized as contributors.

Numerous products today are composites, with components from many different contributors. The common microwave oven in your kitchen has a mechanical body, but the electronics prevail in the eyes of the public. The same is true in modern automobiles. The electronic displays and intelligent screens, back-up cameras, and the like are the focal points. The car's frame, wheels, and body components are a given. Gaskets, seals, and serpentine belts in today's cars are marvels, and yet the public only sees the flashy electronics. The mechanical components are taken for granted and get no attention.

#31

The best designs of mechanical engineers are inherently simple, but simple designs fail to gain recognition.

Should a mechanical engineer come up with a good yet simple design, the public customarily takes the design and invention for granted and does not have a notion of the degree of perfection achieved.

The unsafe tricycle

The safe Big Wheel

One lowly example is that of the Big Wheel®. In the interest of brevity, I will make a simple assertion. The predecessor to the Big Wheel, the common tricycle, is inherently more dangerous for users. Tricycles were designed based on static principles, and yet as dynamic or moving child-ridden toys, the accident and injury rates are high. I have made it my life's work to study and appreciate the many nuances associated with bicycles and tricycles [14].

The design of the Big Wheel is quite advanced over its forerunner (the tricycle), but the public is largely unaware of the differences between them. The Big Wheel offers superior stability over the tricycle. Instead of the tricycle and child tipping, the Big Wheel—with its wider footprint, plastic wheels that slide, and lower mounted rider—tends to slide as opposed to overturning. Unfortunately, mechanical engineers, as many are or become parents, seem blind to the Big Wheel's superior design.

Frankly, the tricycle represents an ergonomic disaster. It is something that should have never been designed or marketed. Yet the tricycle has been around in the same form for well over a century. The same time period ASME has been in existence.

#32

We are put in the situation where apples can be compared with apples instead of apples compared with oranges.

Numerous situations of apples versus oranges occur. Think of a doctor treating a patient. It is very speculative to say what the outcome would have been if another remedy or no remedy would have been used. Lawyers have similar privileges. It is accepted that no two murder trials are exactly the same, and no two juries are going to be identical. Economists can talk all they want, but we have only one U.S. economy at any given time. One can only speculate on what would have occurred if an alternate idea had been implemented.

Another classic case of apples versus oranges concerns the orthodontist profession. Your child has only one set of teeth, and as a parent, it is virtually impossible for you to reject the orthodontist's opinion that your child needs braces. The parent is forced to concede (and pay). The uniqueness of your child combined with the American guilt syndrome preclude you from testing apples with apples. As such, you can only ponder the outcome of the alternate decision for the rest of your life, but you will never know for certain.

The mechanical engineer has no such privilege. Any design proposed and put into practice can be tested or measured against an alternate design because this is the nature of mechanical things. Mechanical engineering devices can be built and duplicated so that an alternative outcome will be known for certain. May heaven help you if you were the designer whose machine or invention came in second. We live in a competitive economic world. Companies that aren't first or as efficient don't stay around long.

#33

Mechanical engineers do not become President of the United States.

Being trained as a mechanical engineer is unlikely to cause one to be a candidate for President of the United States. Of our 45 Presidents to date, I assert that we have had two degreed engineers serve as President. I am editorially removing Washington and Jefferson, since both these early presidents were from an era where generalist training prevailed. In those days, one had to have a wide host of skills.

In terms of modern presidents, the two two engineers, at least by their training or vocation, were Herbert Hoover and Jimmy Carter. Yet history has concluded that both Hoover and Carter were abject failures. Hoover was a geologist by training, studying at Stanford University, but became a mining engineer. He performed fabulously in rebuilding Belgium following World War I and in heading the Marshall Plan to rebuild Europe after the close of World War II, but as President, Hoover was inept. Jimmy Carter graduated from the Annapolis Naval Academy with a general degree in engineering. Carter called himself a nuclear engineer. That description was a stretch as his nuclear credentials amounted to one undergraduate course in nuclear engineering. I will refrain from commenting on Carter's performance as President.

The most common profession to become President is that of law. Lawyers don't solve problems, but rather contain or manage problems. Engineers, however, are trained to solve problems. I once considered a run for the Presidency. Fortunately, I realized I would fail as I would try to solve every problem and wouldn't be able to accommodate the opposition. I assert that any altruistic mechanical engineer would never survive in our political climate. Don't select mechanical engineer as a career choice if you have political ambitions, or you'll engineer your way right out of being elected.

#34

We are viewed by the larger world as nerds.

The world is implicitly aware of the reasons cited above on the issue of whether or not to be a mechanical engineer. As a consequence, any individual choosing to become or to remain a mechanical engineer runs a considerable risk of being regarded as a nerd and a naïve, idealistic fool. Whether a given person actually is one is not the issue. The point is that the person has chosen to be grouped among persons with definite liabilities.

Consequently, our peers, superiors, and inferiors are not as likely to take us seriously. The world tends to put people into categories for the sake of convenience. It naturally follows that mechanical engineers fall into a category of a collection of persons who wear bow ties. It's hard for people to take us seriously with a bow tie on.

Other nerd stereotypes, like using pocket protectors and attending Star Wars conferences, prevail despite the fact the many of us have gone digital (no need to carry pens around) and prefer other TV shows. It doesn't matter that mechanical engineers may be nice people and might not even be nerdy at all—the perception is already there and it's a hard one to break.

But We Are Also Culprits

"Make sure your worst enemy doesn't live between your two ears."
~Laird Hamilton

#35

We are isolated in our profession.

Other professions extend courtesies to fellow members of their line of work. For example, it is common for the families of doctors to be seen by other doctors on a no-charge basis. If a podiatrist's wife is pregnant, in many communities the obstetrician attends to that woman for free. This is a clever way to beat the tax man as it is actually a form of barter income. Doctors are in a high marginal tax bracket. As such, the government would get a generous slice of the money if the two doctors paid each other as normal people (such as mechanical engineers) do. Having babies is a personal expense. The outlay by the podiatrist, if made, would come out of after-tax dollars. Next, the obstetrician doesn't pay tax on income never received. It's a win-win for the doctors, as the tax man has been avoided twice.

Mechanical engineers, on the other hand, just aren't clever enough to figure out this bartering system. Besides, why would one mechanical engineer ever want to consult another one?

Not only do we not extend professional courtesies to other mechanical engineers, we often give away answers for free to friends and neighbors, receiving nothing tangible in return. In addition, if another mechanical engineer actually asked for advice, my experience is that one finds a guarded response. It must be an occupational trait that we are so insecure that we clam up around other mechanical engineers. Helping each other is not our strong suit.

#36

We tend to be cowards.

The collective group of mechanical engineers tends to behave as cowards. Even if the facts are in our favor, the tendency is to acquiesce and allow others who are less qualified to prevail.

As an example, consider legislation that mandates fuel efficiency standards for automobiles. Legislators and environmental advocates influence and drive federal fuel economy standards. The mechanical engineering profession has little say and little pushback. Being pushed around and accepting second class status typifies cowardice.

Our mascot could easily be the Cowardly Lion.

A typical mechanical engineer. (Actually, Bert Lahr as the cowardly lion, 1939.)

#37

The mechanical engineering profession is gutless as per product safety advocacy and standards.

This indictment of the mechanical engineering profession follows directly on the previous discussion of the design of the Big Wheel. The current toy industry code, a set of voluntary compliance standards drafted by the toy industry itself [15], defines a tricycle as a safe toy. This determination was predicated on the fact that the tricycle meets one and only one criterion: a child should be able to place both feet on the ground at any time while sitting on the toy. I believe that the ability for a small child to drag his/her feet during a high-speed downhill runaway hardly characterizes a safe toy. A preschooler cannot be expected to be familiar with Newtonian mechanics of nonholonomic systems. Though were I ever to meet such an extraordinary youngster, he or she would certainly earn my respect.

The mechanical engineering profession has an obligation to act in a responsible manner on behalf of the public and the children who would use these toys. Failure to act in a competent, professional manner to protect the public welfare is regarded as malpractice in most professions.

If we want to claim the privileges and status of being a profession, some consideration should be given to behaving like a profession. A suitable starting place would be to eliminate the tricycle from the marketplace.

It is also worth noting that the public's increased awareness of product safety issues is not due to the efforts of the mechanical engineering profession. Lawyer Ralph Nader, who had no engineering credentials, raised the public's awareness of this in his book *Unsafe at Any Speed* [16].

The inept and negligent pattern of behavior of the mechanical engineering profession is caused by many factors, but among them is the tendency for the mechanical engineer to regard a device by its basic visual appearance (as opposed to the abstract mathematics of the issue) and to predominantly examine one isolated factor or concept without considering how all things work together.

The Chevrolet Corvair automobile had a steering wheel that turned the front wheels. That qualified as adequate from the mechanical engineering perspective. Unfortunately, the Corvair's mechanical engineers failed to note that its steering behavior would become unstable should the driver ever let go of the steering wheel. Even at speeds less than five mph, vehicle crashes occurred. Properly designed steering systems should return to center whenever the driver lets go. Instead of being self-centering, the Corvair's steering was prone to a divergent behavior when the steering wheel was released. Nader was correct in using the phrase, "unsafe at any speed."

When mechanical engineers come up with designs, at times management will overrule various aspects. The engineer has an obligation to protect the greater good. Far too often, we fail to hold our ground. If the mechanical engineer is asked to approve a design that is unsafe, the option of resigning is always present. In my other writings, I have discussed the scandal related to shortcuts in the construction of the World Trade Center's Twin Towers. No one seized the option of resigning in that situation. The supporting steel framing on the Twin Towers lacked thermal insulation application. This oversight—no, the intentional skipping of this step—contributed to the loss of thousands of lives.

One of the reasons electrical engineers are being called upon to work on mechanical engineering problems is that electrical engineers are trained to think in abstract and dynamic terms. The electrical engineer is taught to consider the simultaneous nature of interconnected devices. In contrast, although mechanical engineering paints with a broader brush, the practitioners of mechanical engineering have become myopic. Specialization within mechanical engineering is now our hallmark. The tendency towards myopic vision intensifies as one becomes more and more expert. This explains why those who teach mechanical engineering become intensely specialized. The mechanical engineers responsible for the unstable steering characteristics of the Chevrolet Corvair were myopic and grossly negligent.

The design of the Big Wheel represents, in my view, a nontrivial forward step. However, the failure of the profession to question the obvious erratic behavior and unsafe accident record of tricycles over the last 100 years along with the Big Wheel's 50 years of existence cannot be considered as a plus for the mechanical engineering profession.

#38

Mechanical engineers rent out both our bodies and minds. Moreover, it is hard to turn off the mind—to get it to stop working.

Eric Hoffer [17, page 128] puts it like this: "It may be true that work on the assembly line dulls the faculties and empties the mind, the only cure being fewer hours at higher pay. But, during fifty years work, I can still savor the hours I used to derive working while I was doing dull, repetitive work. I could talk to my partners and compose sentences in the back of my mind, all at the same time. Chances are that had my work been of an absorbing interest I could not have done any thinking and composing on the company's time, or even on my own time after returning from work. People who find dull jobs unendurable are often dull people who do not know what to do with themselves when at leisure. Children and mature people thrive on dull routine, while the adolescent, who has lost the child's capacity for concentration and is without the inner resources of the mature, needs excitement and novelty to stave off boredom."

As the level of mental effort required for the job increases, the employee is less free to own and use his/her own mind for personal pursuits. Furthermore, as the job becomes more taxing mentally, the worker is denied the personal use of his/her mind on the job, and the mind is too taxed to be of full use to its owner at quitting time. If the worker does not feel adequately rewarded on the job from an ego standpoint, then the worker becomes hostile to his employer and the system around him/her. Such is the situation of a vast majority of mechanical engineers.

However, for some (but not usually mechanical engineers) there comes a point when persons who use mental effort on the job become totally immersed in their work and are happy, provided they are relatively immune from criticism and they think they are doing well. They become so immersed that they carry their work home in the form of mental images that continue to be processed even while eating, bathing, cutting the lawn, and so forth. I am describing those individuals who never look at the clock while on the job. If they do look at the clock, it is with the hope that the clock has slowed down. For these people, typified by university professors who don't feel much pressure from the system, they have the luxury and situation to be happy while at work.

Another infinitely happy group is the successful risk-takers who gamble on their own brains and venture out into the business world to be their own person. They rent other peoples' minds. Of those whom the entrepreneur agrees to rent, the good ones move on to a modest reward so as to ensure being kept around, and the not-so-good ones are discarded or driven to a special hell. Somehow this sounds familiar. Haven't we heard lines to this effect before? We mechanical engineers are people who professionally rent our minds.

Another irony is that once a mechanical engineer, always a mechanical engineer. The dissatisfied mechanical engineer is beyond the point of no return. Our education and self-concept deny us the chance to fall back into a routine job such an elevator operator or bowling alley pin-setter. Which is good since those jobs no longer exist.

#39

We tend to be unhappy.

In reading the book *Zen and the Art of Motorcycle Maintenance* by Robert Pirsig [1], I was struck by the fact that the main character was a mechanical engineer taking a motorcycle ride on his vacation. This mechanical engineer had such a complete view of life, yet his view caused him such torment. Surely there must be some connection between the mechanical engineer's personality and the outcome of one's life.

I have searched but have never found a happy mechanical engineer over the age of 40 who is still doing mechanical engineering work. Yes, I have met happy mechanical engineers, but the happy ones are doing other things—sales, management, administration, business ownership. Others have additional credentials such as a law degree, hence involvement in patent law. Mechanical engineers will initially say they are happy, but closer questioning reveals they resent having to park way out in the boonies while others get to park next to the building in the reserved spots. The truth is that mechanical engineers inevitably complain about unfair distribution of workload and preferential treatment of others.

Hippocrates, the Greek father of medicine, put forth a hypothesis that all people may be classified as one of the following:

- Sanguine (fun-loving)
- Phlegmatic (peace-loving and passive)
- Melancholy (perfectionist)
- Choleric (decisive and outspoken leader)

Hippocrates attributed the distinct personality types to innate differences in their blood, as described in the writings of Florence Littauer [18]. The difference in their respective "bloods," so to speak, is deep and permanent.

Whole books have been written on these basic personality traits, various hybrid combinations, and false masks. I will concentrate on the melancholy personality, which has many strengths and weaknesses. The following is an abbreviated list:

Analytical	Enjoys being hurt
Serious	Insecure socially
Talented	Self-centered
Idealistic	Prefers difficult work
Organized	Hard to please
Detail-oriented	Not people-oriented
Makes sketches	Spends too much time planning
Self-sacrificing	Easily depressed
Makes lists	Self-deprecating
Works creatively	Feelings of guilt
Persistent	Hesitates to start a project
Neat and tidy	Unforgiving
Seeks ideal mate	Skeptical of compliments
Economical	Holds back affection
Wishes to please	Has a deep need for approval

It isn't so much the profession itself that makes mechanical engineers unhappy, but those who select mechanical engineering are inherently destined to be unhappy. The unhappiness stems from their basic personality makeup. Much can be accomplished once it is recognized that the composition of a person is largely fixed and unchangeable.

As an empiricist, I do believe a balance of happiness and fulfillment can be found, one where the amount of free thinking allowed versus the level of concentration required on the job corresponds to one's desires. This means that fulfillment and happiness can be achieved, even by melancholies. Even by mechanical engineers.

People who exert minimal mental effort on the job are renting only their physical body to the employer. Their minds are free to scheme, joke, play games, be filled with mischief, or mentally moonlight on the job.

#40

Mechanical engineers tend to be Martha's progeny.

Surely there must be some connection between the ME personality and the outcome of one's life. Recall the story of Martha and Mary in the Bible, Luke 10: 38-42. MEs are Martha's children, and they eventually and inevitably come out of the kitchen as Martha did to complain about unfair distribution of workload.

"As Jesus and his disciples were on their way, he came to a village where a woman named Martha opened her home to them. She had a sister called Mary, who sat at the Lord's feet listening to what he said. But Martha was distracted by all the preparations that had to be made. She came to him (Jesus) and asked 'Lord, don't you care that my sister has left me to do the work by myself? Tell her to help me."

"Martha, Martha," the Lord answered, "you are worried and upset about many things, but only one thing is needed. [For you to keep working in the kitchen.] Mary has chosen what is better and it will not be taken away from her."

We ask for a sour piece of life—working in the kitchen. It is likely that no others, not even the master, will ever come to take their "turn," or to even ask others to relieve us, or to even say thank you. It is unrealistic to think otherwise.

#41

The mechanical engineering profession resembles a bush, and not a majestic tree.

Trees are truly majestic. A strong trunk reaches upward to the sky. Branches spread in an orderly manner to support a canopy of lesser branches and ultimately leaves. A bush, in stark contrast, is characterized by many branches where each branch fights for space and supremacy. At times, the branches in a bush collide. Friction and competition rob the bush of its strength. Bushes can never compete with towering trees.

A majestic tree

A bush whose limbs are clearly competing with each other

As the premier national and international organization for mechanical engineers, ASME has an incredible number of branches. In fact, ASME has approximately 50 technical divisions. Mechanical engineering started out as a split tree, a tree with two major trunks: thermodynamics and machine design. Instead of consolidating its two trunks, the number of competing branches has increased. I say branches as opposed to trunks. No tree can exist, other than the banyan tree, with 50 or more trunks. Mechanical engineering could have at one time evolved to become a banyan tree, but the infighting closed that option.

It's The End of the World As We Know It (or at Least Our Profession)

"Instead of thinking outside the box, get rid of the box."
~Deepak Chopra

#42

Mechanical engineering is a conservative profession.

The mechanical engineer is penalized for taking risks. Any individual who desires to be truly creative and who attempts an intuitive leap is thwarted. I say, "No risk, no gain!" One just cannot stay bottled up and always be 100% correct in speaking and verbalizing if one is to be creative. It is my view that mechanical engineering, contrary to its image, is not a profession that nurtures creativity, because creativity requires trial balloons and an atmosphere of freedom from criticism. But rather, mechanical engineering is proliferated with professional scoffers and cynics.

Criticism of ideas, even trial ideas, is the surest way to stifle creativity, and it is my view that the mechanical engineering profession, as well as the job atmosphere of the typical mechanical engineer, negatively rewards any ill-fated trial idea. Unfortunately, a vast amount of the negativism that we experience comes from within the ranks of fellow mechanical engineers.

We don't have clear limits on our knowledge or area of expertise. There isn't a firm definition of a point at which a mechanical engineer should say "I don't know" or "That question is out of my bounds." This is a serious professional shortcoming. Failure to say "I don't know" creates a situation where the mechanical engineer ventures into too many territories on thin ice. The inevitable consequence is that the mechanical engineer is wrong. Not being taught that it's acceptable to say "I don't know" gives us a serious handicap.

Historically and traditionally, it's often been insecure types that were attracted to mechanical engineering. So much of the mechanical engineer's personality profile can be explained with this simple hypothesis: Mechanical engineers select mechanical engineering because they prefer to work with things and not people. Out of all the choices in engineering, the mechanical engineering curriculum offers—at least at first glance—the safest path. The student thinks mechanical engineering will be easier because he/she can rely on visible perception and reasoning.

This is not only false, but the mechanical engineer has a tendency to shy away from abstract and analytical methods. Electrical engineers are winning the accolades from society today and have largely avoided this trap because electricity can't be seen in the usual sense. Electrical engineers have recognized their blindness, so they've resorted to improving other senses that are significantly more effective at problem-solving in complex situations.

We are ultra-reticent to present our ideas to other mechanical engineers for fear that somebody qualified will shoot down or criticize our design. This, in turn, will damage our ego.

We are hesitant to charge excessively, or at all, for our toil in fear that a high price will cause the potential buyer or user to reject the idea. Rejection is especially hard for mechanical engineering types to take. And I should know—I've had my fair share of it!

#43

Mechanical engineers are no longer at the forefront of

automobile design.

We are no longer qualified to design engines for automobiles and small trucks. Though as a formal profession mechanical engineering is roughly two or three centuries old, compared to the high-tech frenzy, mechanical engineering is old and tired. The real breakthroughs and achievements in mechanical engineering occurred a long time ago.

Technology developments follow what is called the S-curve, with slow growth at first, followed by rapid growth, then reverting to a much smaller rate of growth.

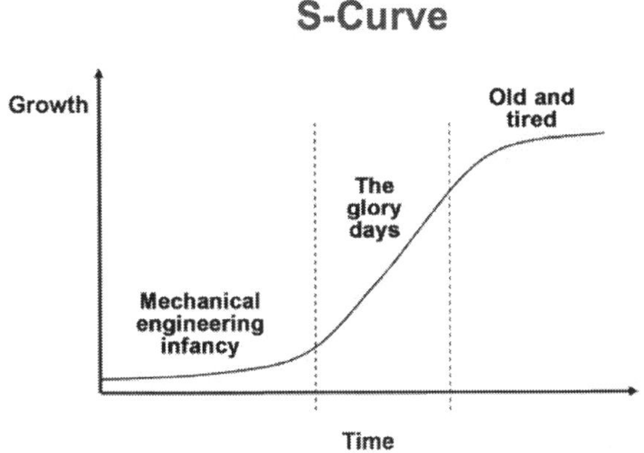

The glory days for mechanical engineering are long past. As the S-curve flattens out over time, the slope is less. It takes considerable effort to achieve even small gains. For example, the development of counter flow heat exchangers was largely accomplished in the first quarter of the twentieth century. The internal combustion engine was essentially perfected from a mechanical engineering standpoint more than 50 years ago. As an example, the 1955 Chevrolet V-8 265 engine was and still is a pretty good engine. In 1957, the Chevrolet 283 cubic inch V-8 developed an incredible 283 horsepower rating. This equaled one horsepower for every cubic inch of engine displacement. In subsequent decades, the focus on better fuel economy caused average horsepower ratings to level off, and even decrease.

Advances in engine design are still taking place, but these are largely related to material improvements, CNC (computer numerical control) machining, and better belts, lubricants, and seals as opposed to fundamental design changes. The changes now being implemented relate to electronics and how engines are managed by computers.

From a mechanical perspective, the changes in engine design are mostly fine-tuning adjustments rather than innovative design breakthroughs. The dramatic breakthroughs in engine technology are largely due to the work of electrical engineers.

Therefore, the future of the automobile and its design are now firmly in the hands of a new team: electrical engineers and their cronies in computer science. The bulk of innovation in vehicles is focused on driving assist, computer diagnostics, and even electrical drive trains. The role of the mechanical engineer has been relegated to second place.

Lastly, the private automobile and its future can be summed up in a few words: self-driving cars. These vehicles will be designed and built by electrical engineers and computer scientists, further discounting the future role of the mechanical engineer.

#44

Mechanical engineers are in the cookie-cutter business.

The mechanical engineer generally designs a thing only once. After we are done with a design, we must move on to a new task or be relieved of a job.

Upon returning to school to complete his MS degree, an engineer with a BS in mechanical engineering and four years of work experience remarked, “A mechanical engineer who is given a problem is expected to have the answer for it immediately as if it (the problem) were trivial… Also, the grunt work given to mechanical engineers in industry can quickly cause the thought process to stagnate."

Our only options seem to be either the impossible or the tedious. Or possibly choosing another career. Which no doubt you are strongly considering if you've made it this far in the book.

#45

Most mechanical engineering has already been done. Moreover, the advances made are usually the result of an elite small circle. Gaining admittance to that elite inner circle is difficult.

David Larsen, a former student of mine, wrote the following note: "Webster defines engineering as the art of creating engines. An engine is something that affects a purpose. Most 'engineers,' however, merely manage another's creation and by this definition are not really engineers. When I started the study of mechanical engineering, my father warned me 'most of the engineering is already done.' I now feel this to be true. As time goes on, I think we will see an increasing pool of generic engineers managing the creations of a small group of real engineers (e.g., Thomas Alva Edison). The only hope for true recognition as a mechanical engineer is to become one of the real engineers. You might as well aspire to be the President of the United States. Your chances of success are about the same."

#46

Textbooks assigned in mechanical engineering curricula tend to be lacking.

The textbooks available to teaching faculty are substandard as contrasted to other professions. Several factors are responsible.

First, there is less motivation for qualified authors to write and publish textbooks in their respective disciplines or specializations. The total enrollment in mechanical engineering stands at about 10,000 graduating BS level students annually.

Not all students take an elective, so let's pencil in a candidate user market at 5,000 students per year. Also, some professors publish notes made available to students. Hence, not all students have an assigned book. Textbooks get recycled, as the campus book stores buy back used books and resell them. The author gets paid a royalty only once—when the book is sold the first time. In any topic or specialty, there are competing authors. No single author will corner the market. Even if an author is successful, the best he/she can expect is 20% of the market, hence 500 or possibly 1,000 books annually. Many schools, especially smaller enrollment schools, struggle to maintain their accreditation. A common practice is that smaller schools will want to appear cutting-edge. Any mechanical engineering text is likely to be dropped when it becomes five years out of publication date. The likelihood of selling only 2,500 or even 5,000 copies of a work, a work that might have taken several years to produce, provides little incentive for authorship.

A second factor relates to promotion within academia. In my case as a faculty member, one message was clear: The authorship of a textbook was viewed negatively by my department head and the Dean of Engineering. Their priority was papers published in scholarly peer-reviewed journals along with writing grant proposals and obtaining external funding. Authorship of a textbook stood as *prima facie* evidence that the faculty member had spent his/her time in unwanted and counter-productive pursuits.

The small enrollments in mechanical engineering as compared to, say, microbiology, law, mathematics, philosophy, history, and humanities assures the mechanical engineering textbooks are wanting.

#47

Mechanical engineering BS-level graduates are not ready to be productive employees.

Mechanical engineering is such a broad and encompassing field that it can't (speaking for most students) be digested and assimilated in four years of study. The idea to expand to a five-year curriculum was rejected based on the economics. Mechanical engineering had to stay at four years or lose enrollment.

The reality is that the average BS graduate lacks the tools, knowledge, and world view to be productive. Employers recognize that the BSME graduate is not ready to be a productive employee.

#48

We believe in and champion the doomsday end of life.

Mechanical engineers consider the Second Law of Thermodynamics the bedrock of the universe.

This fundamental tenet of the mechanical engineering profession states that, when the universe is considered as a closed system, it will approach a constant temperature. Absent temperature differentials, there can be no useful work, referred to as Gibb's free energy. Life will cease to exist. When everything in existence reaches a uniform temperature, engines that produce useful work will cease to function.

Mechanical engineers take the Second Law of Thermodynamics as sacrosanct. To us, it is everything. No life or activity of any kind will be possible due to this lack of available energy in the sense of J. Willard Gibbs (1839-1903). Yet the rest of the world doesn't share this view. Stars will continue to exist. Life will go on.

The Second Law of Thermodynamics is seriously flawed as those who choose to expound on it ignore the reality of the internal creation of information and the consequence thereof. Certain literature argues that the creation of information requires energy and that the Second Law is not really violated; however, that literature deals only with particles at the molecular level and not with aggregates of particles commonly called objects. While the jury is still out on this issue, it is my view that given information, which can be created at a minimal energy expense within a closed system, the entropy of a closed system can decrease of its own accord. The thought experiment "Maxwell's demon" suggests this is possible; specifically it posits that the Second Law of Thermodynamics could hypothetically be violated. In this thought experiment, a "demon" controls a door separating two chambers of gas, and he opens the door when fast molecules are near it so that the fast molecules move into one chamber while the slow molecules move into the other. This creates a situation where we have one hot chamber and one cold chamber, and thus the demon's behavior has decreased entropy, which violates the Second Law of Thermodynamics. We do live in the age of information, but the mechanical engineering thermodynamic types have chosen to ignore such a well-known and publicized fact. One could argue there is more than one demon in this scenario, with the mechanical engineering profession being a demon all its own!

Any profession that fails to modify the statement and applications of the Second Law of Thermodynamics as it is now taught in the mechanical engineering curriculum isn't a group worthy enough to be taken seriously. The Second Law of Thermodynamics is still the subject of debate in the literature. It is true that thermodynamics experts remain smug and confident owing to the continued absence of a valid counterexample, but it must be recalled that thermodynamicists have never supplied proof that their Second Law is correct or valid and that the world of common people have formed a view (be it right or wrong) that information costs little and that information can bring order to a situation where chaos previously prevailed.

#49

In recent history, mechanical engineers have been known

as losers.

For the last 75 years, mechanical engineers have borne a legacy of losing. We had our heyday from the mid-1700s to the mid-1900s, but in the middle of the twentieth century, the tide turned. World War II was the beginning of the end. Mechanical engineers were ill-prepared to handle the challenges imposed by the war. The great breakthroughs that were achieved and that won the war for the Allies were largely due to the work of electrical engineers, mathematicians, and physicists. Major results were the development of radar, inertial navigation, guidance and control systems, LORAN, information systems, code breaking, and the Manhattan Project.

A significant reason the Allies won can be attributed to a single man: Dr. Vannevar Bush [19]. He was a high-ranking scientist, an administrator in the Department of Defense, and an electrical engineering graduate from M.I.T. Bush personally put together a team of top-level electrical engineers, physicists, and mathematicians at M.I.T. Mechanical engineers were excluded for all practical purposes. A dramatic monument to the achievements of the electrical engineers in World War II is the 28-volume Lincoln Laboratory Radiation series of books published by McGraw-Hill at the close of World War II. This series declassified a large portion of the electrical engineering advances made during World War II. It is true that mechanical engineering concepts such as kinematics are contained in one of the volumes, but the overwhelming emphasis is on electrical engineering. The Lincoln Laboratory Radiation series rang the final bell in mechanical engineers' fight for a position as a premier engineering degree.

Other setbacks for mechanical engineers since World War II include Sputnik, the economic obsolescence of the Sparrows Point Steel Plant near Baltimore, the development of the Linz-Donawitz steel-making process in Austria in 1963, and Thomas J. Watson's first commercial digital computer released in 1955.

Each of these events hastened the decline of mechanical engineering by shifting the focus of American research and manufacturing from mechanical machines to electronics and computer-based development.

Today it is commonly accepted that mechanical engineering industries are on the wane. The current financial difficulties of many once-powerful companies of America now bear this out.

#50

We restrict our field of practice.

The field of mechanical engineering looks inward and builds boundaries around itself. Charles Babbage (1791-1871), a French philosopher and mechanical engineer, is credited by many as the inventor of digital computing. Sadly, the high and almighty priests of mechanical engineering, notably thermodynamic purists, erected a barrier that placed digital computing outside of the fort's perimeter.

#51

Although mechanical engineering is inherently an art, the teaching of the art is largely absent.

The teaching of mechanical engineering is done by specialists, hence insects. The varied specialists regurgitate to students the minutia of the multitude of sub-parts. Mechanical engineering students get hammered with science and analysis, and only get lip service to the art of mechanical engineering. Students are trained to turn the crank, to crank out answers. Paralysis by analysis becomes common as creativity is stifled. In true art there is no predetermined "correct" answer.

#52

The mechanical engineering profession has thrown the baby out with the bathwater.

The old-timers were not idiots. Without today's modern

computers and analysis tools, they used ingenuity combined with empiricism to solve problems. With the advent of modern digital computing, the newer generations look to computing to get answers, becoming dependent on them. When that trait is added to reliance on memorization, the result can aptly be described as paralysis by analysis.

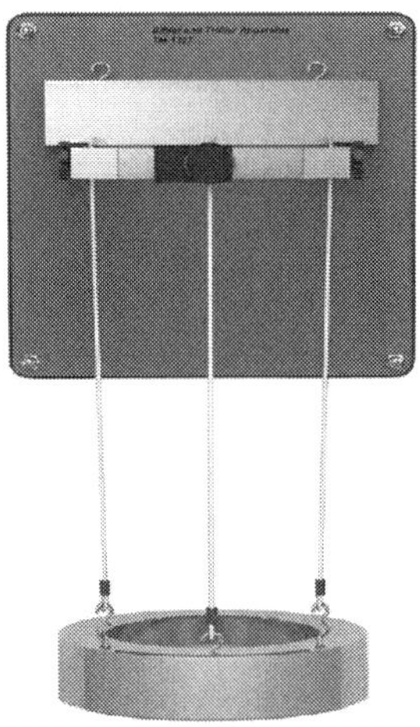

The illustration depicts an experimental apparatus commonly referred to as a trifilar pendulum. The pendulum and its usage will be discussed below.

The techniques used by the old-timers have largely been forgotten. This loss is unfortunate, and I will cite just a few illustrative examples of the many lost skills.

Assume that an engineer needs to know the mass moment of inertia of some object or assembly. The old-timers determined the mass moment of inertia by placing the object on a platform suspended by three wires or strings. This device was referred to as a trifilar pendulum. Gravity was known, as was the length and configuration of the strings. The mass and center of gravity were also known, based on simple measurements. The object resting on the platform was twisted slightly and released. The period of torsional oscillation was measured; thus allowing the calculation of the object's mass moment of inertia. This calculation was done using a device called a slide rule. Today's modern high-tech mechanical engineers will, in contrast, spend an inordinate amount of time staring at a screen and typing to come up with an answer using analysis. The old-timers had the answer more quickly and with greater accuracy.

An engineer desires to know the sweet spot of some object. Think of a baseball bat or tennis racket. The sweet spot is defined as the point where there is no reaction or "sting" at the object's handle or pivot point. The sweet spot is also known as the center of percussion, also defined as the radius of gyration. The old-timer would suspend the object on a string attached at the handle or pivot point. By striking the object at varied points, the center of percussion was determined when the point of suspension did not shift when the object was struck. The old-timer had the answer quickly. Today's newer generation of engineers again becomes fixated staring at a computer screen.

Early engineers had a thing called a free-surface water table. This was used for visualization of two-dimensional fluid flow. The water table could be adjusted to produce visualizations for both subsonic and supersonic air and gas flows. In the 1980s at the University of Illinois, the free-surface water table was scrapped to make room for more computer stations. Today's high-tech mechanical engineering students spend their time staring at computer screens.

In olden days, it was common to want to know the area enclosed within some closed curve. One typical application came from the pressure-volume indicator curve for a steam engine. The P-V diagram was recorded on a paper chart. The old-timers had a device called a planimeter, which had wheels and screws that would record the integral of the movement of a tracing. The planimeter gave results quickly and within two digits of accuracy. The old-timers also had alternative approaches. For example, they would cut out a shape from a material of known thickness and weight per square unit of area, and obtain the answer by weighing the cut-out shape. Of course, today's high-tech geniuses resort to one thing—a computer—to perform an analysis.

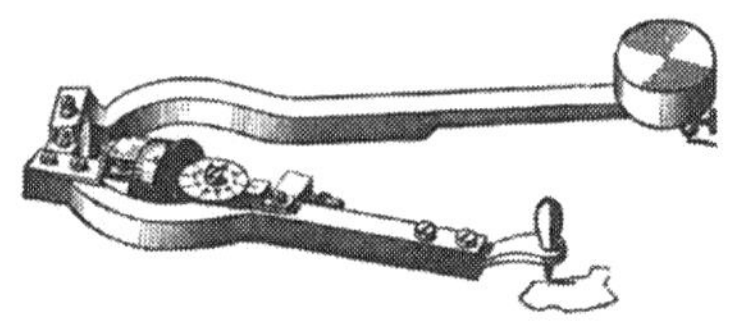

The mechanical integrating device above was designed to measure the area within a closed curve. The stylus was moved along the curve's path, coming back to the starting point. Based on geometry and the ability of a lead-screw to accumulate angle, a dial would provide the answer: the desired area within the prescribed enclosure.

One final example of a lost skill: The Buckingham Pi Theorem. This theorem is used to derive dimensionless numbers such as the Reynolds number, the Froude number, and the pressure coefficient, which are essential to fluid mechanics. Few mechanical engineering students are even exposed to the Buckingham Pi Theorem anymore, much less appreciate its use.

I say the baby has been thrown out with the bathwater. The old-timers had incredible techniques for solving problems, using their creativity and imagination rather than relying on computers to do their thinking for them. If they continue throwing out all these babies, who will be left to carry on the mechanical engineering profession?

#53

Problem-solving in mechanical engineering has been superseded by drudgery and dogma.

When I taught mechanical engineering, at times students would become upset that somebody got an answer, such as on an examination, faster than them. The implication was that it was unfair that somebody else used a trick to achieve a solution.

My philosophical view was that the goal of mechanical engineering was and is to obtain solutions as efficiently as possible, with as little effort as possible. The objective is to find the most direct method, and the list of direct methods has no bound. As a professor, I could never tabulate a listing with instructions for every conceivable shortcut. If some student, say, on an examination, discovered and identified a new shortcut, all the better.

Mechanical engineering is and remains an art form. Creativity can't be stifled, or at least it shouldn't be. I fear that the modern-day insects teaching mechanical engineering don't share or appreciate my views.

One of my former students, who now has his doctorate, describes what he sees in mechanical engineering education as "The Asianization of Education." The inference is that a professor who is from Asia or elsewhere outside of the U.S. and speaks with a foreign accent is somehow held up as the pinnacle in education. I am opposed to the presumption that simply having a foreign accent makes a person intellectually superior. Additionally, we should encourage creativity within our own educational system so we don't fall behind other countries.

#54

We tend to be perfectionists.

I make this note merely as an observation. Perfectionism is a great trait, but it has its price. The mechanical engineer has to be on guard to prevent becoming preoccupied with some facet or aspect of a design to the exclusion of all else.

But Don't Forget to Laugh

"If you can laugh at yourself, you are going to be fine. If you allow others to laugh with you, you will be great."
~Martin Niemoller

#55

Mechanical engineering is circling the drain.

The water is swirling, and mechanical engineering is about to go down the drain. The profession is dying for many reasons that are intertwined in the statements, reasons, and explanations appearing in this book. It would be appropriate, however, to review the major reasons.

First, we have failed because our language skills are at a remedial level. Specifically, our vocabulary is poor and lacks precision. Too many ambiguities exist. The problems coming along the pike are too abstract to be handled by persons with an inadequate and improper vocabulary. For example, since when does a radiator radiate? Why have mechanical engineers permitted the radiator to be called a radiator for so long when it is, in fact, a convector? This demonstrates our failure to be definitive and assertive in our language. The radiator vs. convector usage also proves that mechanical engineers, as a profession, allow faulty thinking to prevail. Most mechanical engineers have never given the radiator in their vehicles any serious thought.

Second, with the proliferation of community colleges and opportunities for online self-education, multitudes of entry-level technical people are now performing work previously performed by degreed mechanical engineers. The requirement to have Professional Engineer designation is no longer adhered to.

Third, we fail to repel and correct misconceptions. Radio broadcasters and meteorologists make incredibly absurd statements. One recent broadcaster, during a cold wave, asserted that breathing air at -70 degrees F wind chill would freeze one's lungs. In reality, once air is inhaled, the air is at its ambient temperature, and even rising, as the human lung has a high surface-to-volume ratio. The assertions made by the media about wind chill are commonly blatantly erroneous. What would happen if newscasters made erroneous statements about law and medicine? Those professions would scream and demand an immediate retraction and correction. But not the meek mechanical engineer.

Fourth, along with an imprecise language comes an imprecise mindset. Before Newton's time, the problems encountered by the scientific community defied clear explanation because the abstract language of mathematics was inadequate. Once Newton had the language clearly defined, he was able to slay problems left and right. It is interesting to note that Newton had no intention of revealing his language and thinking tools except in a posthumous volume. He had the right idea—that a profession requires secrecy and a closed membership list. It turned out that his results were so sensational in numerous areas such as optics, mechanics, fluids, and so forth, that he was besieged by a close friend who happened to come across a strange manuscript in Newton's living quarters. This manuscript revealed Newton's secret language—The Calculus. Newton's friend, Sir Edmond Halley, persuaded Newton to publish *Principia Mathematica*, considered to be one of the most important works in the history of science, leading to a great revolution in physics.

Mechanical engineers have not used language to keep the

club exclusive and private. Instead, we have lowered admission standards. The mechanical engineering profession has allowed its words to be used by pretty much anybody. Think of the words *screen, filter, bolt, nut, jerk, slug, heat, hard, bearing, lever, valve, fuel, ash, fan,* and *duct*. There is nothing exclusive in these words. Virtually anybody can use them fluently. In reflection, the degradation of the language and the profession has been accelerated by the corruption and misuse of the mechanical engineers' working vocabulary.

A proposed remedy would be to adopt a new language of some obscure islanders. All technical words could be coined as tongue twisters. Electrical engineers who study system theory have taken this approach, using the languages of the remote island tribes of Banach, Sobolev, and Lebesgue. If anyone doubts the truth of this, try to scan a recent issue of IEEE *Transactions of Automatic Control.*

Mechanical engineers use the word *jerk* to denote a scientific, precise, and technical thing. Any professional group who would coin and continue to seriously use that word as a technical term shouldn't be taken seriously. If jerk isn't enough for a questionable status, consider the words slug, nut, and screw. Whatever happened to snail and acorn? We also talk about the power of a horse and tons of air conditioning. Surely we could come up with more sophisticated and impressive terms.

Mechanical engineers are blind to the events of the world. We have continued to toil in a heads-down fashion, ignoring the great chances for stardom which often occur at the intersection of fields such as mechanical actuators. Other professions have spotted these opportunities and will continue to exploit them. For example, the IEEE society sponsored conferences on Mechatronics. The sponsors first coined the word *mechatronics* to mean the interface of mechanical actuators and electronics. It is interesting that half of the new thing is mechanical, but the mechanical engineering profession was largely absent.

#56

Our children may want to follow in our footsteps.

If you become a mechanical engineer and have children, they just might look up to you, due to innocence, and might try to follow in your footsteps. But if you look ahead to your old age, doesn't it seem logical to guide your children into professions in which they will be better off? And, more importantly, financially capable of supporting you as you get along in life? Those senior living homes aren't cheap!

#57

Mechanical engineers are lacking in sense of humor.

Finally, it's important in life to be able to laugh at yourself. I wrote this guide with humor in mind. However, I have yet to hear much laughter from mechanical engineers.

Who Agrees With Me?

After drafting this book in its early form, I decided I needed some perspective, particularly for the long negative list of snares, pitfalls, and dead-ends. Of dozens of reviewers, not a single person challenged my statements. Moreover, when viewed as a whole, the listed items became synergistic. They fed and built upon each other. In concert, the fifty-plus negative reasons combined to achieve a weight greater than the sum of the individual items.

One undergraduate student who read an early draft became discouraged. He asked me why I wrote such a depressing and devastating document. Yes, I acknowledge it has depressing aspects. I answered the student's question with another question:

"Why did Newton invent the calculus?"

In short, Newton was faced with a world of confusion and unanswered questions. The calculus permitted Newton to address the confusion and develop answers. Likewise, I wrote this book to foster understanding and order regarding the world of mechanical engineering.

A definite anti-mechanical engineering sentiment exists in the marketplace and in the minds of the public. Consequently, the majority of persons who reviewed the early drafts of this material agree enthusiastically. Yet other groups reacted differently.

Many of my former mechanical engineering students concur with my sentiments wholeheartedly. They readily admitted that mechanical engineering is in dire straits in terms of image. They stated that the workplace situation in which mechanical engineers find themselves has drastically—and negatively—changed in recent years. The young mechanical engineers have to fight their nerd image as well as the perception of being dinosaurs and historical losers. The predominant notion voiced was that they had to continually justify their existence and that they should be considered as part of the new solution rather than viewed as part of the old problem.

Traditional mechanical engineering professors, who are the mainstream of mechanical engineering education, found this material off-putting. Two administrators even returned the material, only partially read, stating that they "might consider" continuing reading if and when I provided an adequate number of positive reasons for selecting mechanical engineering as a professional career choice. One fellow professor, a young man, said he found the material to be "offensive" and that he "didn't like it." My view is that we should deal with reality as it is. The issue of like and dislike is quite immaterial.

The reaction of juniors and seniors in my undergraduate mechanical engineering program was quite different. Young people are quite impressionable, and this type of material can be unsettling to a fresh, naïve mind. Undergraduate students often asked me, "Why did you write this? Aren't you one of them?" Others stated, "The document didn't deter me from becoming a mechanical engineer, but rather it strengthened my resolve."

One mechanical engineering student, in reaction to reading my draft, wrote a note stating: "I had not realized how very little I really know about what I'm getting into; this is disheartening. I suspect that few of the people you inform on this topic can believe you, let alone accept the devastating verdict(s) you have set forth."

Many other students and former students in mechanical engineering have expressed the fact that, if they had it to do again, they would pick something with more of a high-tech image, notably computers and electrical engineering. One of the former students who holds both a BS and an MS in mechanical engineering stated that he selected mechanical engineering because he believed it would be easier to deal with mechanical engineering systems since they can be visualized. He then realized that his line of thinking was a trap as "things are not what they seem." It is fair to say, however, that most students are overwhelmed by the list of reasons above and are unable to comment intelligently.

While juniors are in a daze, seniors are more practical as they consider ways to appear high-tech by joining nontraditional mechanical engineering societies such as ACM, SIAM, AMS, IEEE, and ORSA. Their other course of action is to select technical electives with greater emphasis on computer sciences, electrical engineering, advanced mathematics, CAD-CAM, finite element analysis, control systems, and Operations Research.

Professors in other disciplines expressed extreme interest. Electrical engineers and computer scientists had difficulty believing that most mechanical engineers are largely unaware of the existence and impact of the Kalman-Bucy paper [4]. It is not an unreasonable guess that the Kalman-Bucy paper, as measured by the Citations Index, would outperform the remaining top five or ten most cited ASME papers combined (excluding other R. E. Kalman papers, of course). The citation count of that paper grows at a rate of several each day. No other paper published in *ASME Transactions* can even come close, and the citations just keep coming. One agricultural engineering professor was most interested in the list of negative reasons. He was fully aware of the serious trouble that agricultural engineering was in, and he hoped to use the material to better understand and deal with agricultural engineering's current problems because of the close resemblance to mechanical engineering.

I shared my thoughts on mechanical engineering with interviewers for two large Midwestern companies, who are responsible for hiring engineers, math majors, and computer scientists. They were very interested in gaining insight regarding today's engineering students. The interviewing companies expressed a common concern that even the high-GPA student in today's college recruiting scene is a dismal prospect with respect to creative problem-solving, adaptability, and plain common sense. The students that they see are filled with disorganized and memorized facts, and they only have a minimal ability to understand, use, and creatively apply the bulk of that collection of factual material.

If problems ever exist in any endeavor in life, the best hope for solving them arises when the problems are identified and scrutinized. It has been my thrust to pinpoint and highlight problems associated with the mechanical engineering profession and to make the potential entrant aware of these problems.

The mechanical engineering profession has had some severe image and public relations issues. Notably, mechanical engineering is not on the high-tech bandwagon. But these problems are not of sufficient magnitude to deter the properly motivated student.

The mechanical engineer and the mechanical engineering profession should address the issues of risk and aggression. A person of the world must have aggression and take risks in life. Much of the world is dictated by illusion as individuals, groups, and nations act as adversaries and rivals. Whether it be in street fighting or nuclear war, the perceived best way to get through is to be a tough guy with weapons and a will to use them. This means fighting tough and taking risks.

If mechanical engineering is to be viable as a profession of the future, those in it should seriously consider staking claim to some defined territories and acting aggressively so as to establish said territories as belonging to mechanical engineering. The objective is not necessarily to lay claim to any old territory but rather key territories such as the interfacing of mechanical systems and information systems.

In our current high-tech society of experts, critics, and instant replays, there is a tendency to avoid risk. My personal philosophy is that the greatest risk is to avoid risks.

How to Revive Mechanical Engineering

Practitioners in most professions have a natural desire to be treated with respect and dignity, but that respect and dignity must be earned. The mechanical engineering profession has attempted to cleanse its genes by selective culling. The high priests see that science, rather than art, is the ticket to a new being. The concept of mechanical engineering as an art form is being banished. Those who profess that mechanical engineering is a creative art form are being systematically drummed out.

Personally, I lament for mechanical engineering as a profession. I entered academia half a century ago. I had great expectations. I even had a wonderful and rewarding professional life. Sadly, if I were entering the field today, I can state two things to be true:

- The powers that be wouldn't have me, and
- I wouldn't have them.

Evolution can be painful. The genetic cleansing that mechanical engineering is undergoing is especially severe.

If nothing is done to remedy the problems and obstacles facing mechanical engineering, the profession will still survive, but only as a collection of specialties. Typical specialties will include heat transfer, materials, CAD-CAM, artificial intelligence, nano-engineering, bio-medical applications, and heating/ventilation/cooling. It used to be said that a mechanical engineer was the "engineer's engineer," meaning that the mechanical engineer was the real thing when the going got tough. It is probable that electrical engineering has eclipsed mechanical engineering to be the pinnacle of true engineering. The old days of mechanical engineering consisting of half thermodynamics and half machine design are now just a memory.

Mechanical engineering had its origins in the steam engine. Its two parents were thermodynamics and machine design. Over the years and even centuries, new shoots branched out. Today, ASME has approximately fifty distinct shoots or branches. Instead of being a tall and majestic tree, mechanical engineering has evolved into a bush. Each stem fights to be the supreme or center one. As is the case with bushes, the many branches are intertwined. They fight and rub against each other, and the in-fighting is intense. Harmony does not prevail.

Further reflection suggests that the problems facing the mechanical engineering profession are complex, deeply rooted, and not amenable to an easy repair or facelift. As a consequence, there is little that the mechanical engineering profession, or the mechanical engineer as an individual, can do to turn the situation around. Meanwhile, the high-tech snowball of electrical, information, and computer technology gathers steam and advances on. It's interesting that the phrase "gathers steam" stems from the nearly forgotten mechanical engineering heydays of steam locomotives and steam engines.

When the world changes, the specialists and insects perish. The University of Illinois once had a proud and strong Railway Engineering program circa 1920. Today that program has vanished except for a few reminders such as a building that is still called the Transportation Building. Railway engineering at Illinois died because the faculty refused to accept diesel-electric engines and instead insisted that steam was king. Ironically, other schools have now capitalized on transportation, notably M.I.T. and the University of Tokyo.

Companies, too, can die as a result of refusing to adapt to new ideas. Take the National Cash Register Company (NCR), who believed in mechanical cash registers in the 1920s. They believed so strongly that they laid off an employee by the name of Thomas J. Watson, who had different ideas. Watson went on to found IBM.

Reflections on the Mechanical Engineering Curriculum

How can a student enrolled in mechanical engineering get more bang for the buck? The question ultimately boils down to which elective courses are selected.

At the University of Illinois, the question of generalization versus specialization was always present. Students are required to take courses both within and outside of the mechanical engineering curriculum. The mechanical engineering curriculum is heavily dominated by required courses. Some elective options do exist, usually taken in the senior year after most required courses have been taken. Of the elective credits available, the decision is critical. How does a student choose wisely?

An all-too-common choice is to take a blow-off course just to get an easy A. I strongly advise against such an approach. The modern university has far too many good courses and good professors to ever justify taking a blow-off course. I once had a male student tell me he took Basket Weaving 101 just to meet attractive girls. My reaction was that other means on a campus exist to achieve that objective. One needs to keep one's priorities straight in life. In cases when students were unsure of their future professional directions, I consistently relied on one recommendation: take more courses in mathematics, particularly advanced calculus.

Beyond satisfying the requirements for a BS in mechanical engineering, use the available elective courses in the areas within and without mechanical engineering to unofficially satisfy the requirements of an *arête* option. At present, this option remains unofficial as students and their advisors have the ultimate say in how to use their elective hours.

The *arête* option consists of the selection of *arête* curriculum electives. The premise is that students can strengthen their professional careers based on wise use of the available options.

The political task of convincing an entrenched bureaucracy of the merits of the *arête* option is not worth the trouble. Because the students enrolled in mechanical engineering are satisfying the full and normal requirements for a degree in mechanical engineering, there is no formal requirement that the faculty and administration approve of the option. All students who choose to affiliate with the *arête* option do so on their own. Even the issue of admission would be strictly an informal and individual decision.

Generalists can be identified by their behavior patterns. Generalists seek challenges and opportunities to learn in many broad areas in spite of hardships and perhaps lower grades. Generalists also feel compelled to focus on their weaknesses with the goal of improvement, while specialists avoid new areas and try to take the easy and comfortable route. Specialists can be identified then by their behavior, which says, "How little do I need to do in order to get by (with respect to educational matters)?"

The student's grades may get him or her the interview, but after the interview starts, what is important is what is said. Interviewers are keenly looking for spark and initiative. A transcript with blow-off courses becomes a turn-off. In contrast, a tough course, even with a poor grade, becomes a sign of a student seeking challenge and an expanded mind. Consider taking, for example, a linear algebra course for math majors—as opposed to the watered down linear algebra course for weaklings. The interviewer will recognize aggressive pursuit of knowledge.

As I look back, my biggest mistake was in not taking the time to understand how to study. I strongly advise that students form a study group. My second biggest mistake was in not selecting my friends wisely. Just because some computer assigns you a roommate in college, don't assume that roommate will become your best friend. You can be civil and polite, but be careful about becoming too close to a stranger without scanning for characteristics.

This book can be viewed as a pile of negativism, or conversely as a goldmine of information. I view it as the latter. If I had to do it all over again, I would again select mechanical engineering as my major. But maybe a different roommate. And with any luck, I'd see you in Advanced Level Differential Equations. Maybe we'd even form a study group.

Closure

As stated at the onset, much of this book was drafted some 35 years ago—in 1984. Now that I am retired, I can take the time to dust off and return to earlier writings. With the passing of those 35 years, the world has seen vast changes. One notable change has been the advances made in the technology of word processing. In the early days I indeed did cutting and pasting. Bill Gates became wealthy, but in my view he deserves every penny. Modern word processing has been a blessing for me.

Despite the acclaim that the computer has received, other things impacted my life and in turn my productivity. My education as a mechanical engineer has had a profound impact on my life. I want to quote a wise man, Dr. Doug Marriott: "An education is what you have after you have forgotten everything you learned in school."

I feel immensely educated. I feel that I have been incredibly blessed. My education as a mechanical engineer, and thus as a generalist, has permitted me to attack and slay problems at will. As problems have come into focus, my mind has provided answers. The list of what I call my solved problems is long. I answered the bicycle stability problem that evaded a solution for over a century. The iCan Bike program, my brainchild, has permitted tens of thousands of children with disabilities to master bike riding. I have answered the question of ice age causation. I have provided firm insights into the question of global warming. I have explained why the World Trade Center Twin Towers collapsed as opposed to standing after being impacted by commandeered aircraft. I have debunked the legend that lemmings rush to the sea and commit suicide. These undertakings are detailed in my forthcoming autobiography, *Dumb Dickie* [20].

The doubters and skeptics will argue with me and my boastings. My objective has never been to convince all. I again paraphrase Patton, “It is only important that I know.” When I know the answers, I can be content and move forward.

Mechanical engineering represents an educational treasure. The student becomes equipped in myriads of ways. I strongly endorse such a choice. To those who have been alarmed at these writings, perhaps mechanical engineering isn’t for you. As a professor, I strongly encouraged faltering students to reexamine the world and their many options. Life is far too wonderful to become trapped doing something that isn’t your thing.

Peace, blessings, and happy trails to you.

Acknowledgements

I am indebted to great thinkers, professors, and colleagues. These include Dr. R. E. "Gene" Goodson, Dr. Douglas L. Marriott, Dr. Donald Olson, and Dr. Branny *von* Turkovich. I also feel blessed to have had associations with my many students who had such bright minds.

About the Author

Richard E. Klein, by his own admission, is an incurable romantic and altruist. His writings and musings are filled with hope and bright horizons despite having lived through World War II and the Korean War as a child, both of which deeply impacted his worldview. Through his books, he aims to point the way towards a better internal mindset and a better world.

Richard earned his Ph.D. in engineering from Purdue University in 1969 and taught systems theory for three decades at the University of Illinois in Urbana-Champaign before retiring in 1998. He holds a particular interest in bicycle stability and control, and has devoted much of his time and energy to the development of an international program for teaching children with disabilities to master bike riding. Visit iCanBike.org and RainbowTrainers.com for more info.

Richard and his wife of more than 50 years, Marjorie Maxwell Klein, reside in the St. Louis area. They have two children and six grandchildren. Richard writes for them and for generations to come.

We're All Set, *Kisses When I Get Home*, and *The Deadly Gamble* are some of Richard's books currently available, with many more in various stages of writing and publication.

References

1. Pirzig, Robert M., *Zen and the Art of Motorcycle Maintenance: An Inquiry Into Values*, Morrow, New York, 1974.
2. Heinlein, Robert A., *Time Enough for Love*, originally published by G.P. Putnam's Sons, 1973, reissued by Ace Books, 1987. Page numbers reference 1987 Ace Books printing.
3. Karman, Theodore von, *The Wind and Beyond: Theodore von Karman, Pioneer in Aviation and Pathfinder in Space*, Little, Brown and Company, 1967.
4. Kalman, R.E. and R.S. Bucy, "New Results in Linear Filtering and Prediction Theory," *ASME Transactions of Basic Engineering*, Vol. 83(1), pp. 85-108, March, 1961.
5. Iacocca, L.A., *Iococca: An Autobiography*, Bantam, New York, 1986.
6. Kalman, R.E. "On the General Theory of Control Systems," *Proceedings of the First Congress, International Federation of Automatic Control*, Moscow, 1960, published by Butterworths, C.F. Coales, editor, London, pp. 481-492.
7. Jones, D.E.H., "The Stability of the Bicycle," *Physics Today*, Vol. 23 (4), April, 1970.
8. Roland, Jr., R. D. and Massing, D. E., "A digital computer simulation of bicycle dynamics," *Cornell Aeronautical Laboratory Inc Technical Report No YA-3063-K-1*, June 1971.
9. Astrom, K.J., R.E. Klein, and A. Lennartsson, "Bicycle Dynamics and Control," *IEEE Control Systems Magazine*, Vol. 25(4), Aug. 2005, pp. 26-47.
10. The Mark 14 Torpedo [Wikipedia]. <https://en.wikipedia.org/wiki/Mark_14_torpedo>
11. Glueckstein, Fred, *Churchill as Bricklayer*, Finest Hour 157, Winter 2012-13, p. 32. <https://winstonchurchill.org/publications/finest-hour/finest-

hour-157/churchill-as-bricklayer/>

12. "Hoverboard inventor says he has made no money - mostly because of cheap Chinese knock-offs," *The Guardian*, 10 January 2016. <https://www.scmp.com/news/world/article/1899720/hoverboard-inventor-says-he-has-made-no-money-mostly-because-cheap>
13. Welch, Ashley, "Thousands of kids injured by hoverboards in their first 2 years on the market," *CBS News*, March 26, 2018. <https://www.cbsnews.com/news/hoverboard-injuries-26000-children-hurt-in-first-2-years-of-sale/>
14. Klein, R.E., *As Easy as Riding a Bike*, Independently Published, in press, to be available from www.amazon.com, pending in 2019.
15. The Toy Association, U.S. Safety Standards, standard last revised 2017. <http://www.toyassociation.org/ta/advocacy/federal/standards/toys/advocacy/federal/us-safety-standards.aspx>.
16. Nader, R., *Unsafe at Any Speed*, Richard Grossman, 1965.
17. Hoffer, E., *Between the Devil and the Dragon*, HarperCollins Publishers, 1982.
18. Littauer, F., "Chi Alpha Discipleship Tool," from *Personality Plus: How to Understand Others by Understanding Yourself*, Fleming H. Revell, 1992. < https://irp-cdn.multiscreensite.com/2988a589/files/uploaded/personality-plus.pdf>
19. Bush, V., *Science: The Endless Frontier*, ACLS History E-Book Project, 1999
20. Klein, R.E., *Dumb Dickie*, publication pending.

Made in the USA
Middletown, DE
25 May 2019